Symmetry and Heterogeneity in High Temperature Superconductors

NATO Science Series

A Series presenting the results of scientific meetings supported under the NATO Science Programme.

The Series is published by IOS Press, Amsterdam, and Springer (formerly Kluwer Academic Publishers) in conjunction with the NATO Public Diplomacy Division

Sub-Series

I. Life and Behavioural Sciences	IOS Press
II. Mathematics, Physics and Chemistry	Springer (formerly Kluwer Academic Publishers)
III. Computer and Systems Science	IOS Press
IV. Earth and Environmental Sciences	Springer (formerly Kluwer Academic Publishers)

The NATO Science Series continues the series of books published formerly as the NATO ASI Series.

The NATO Science Programme offers support for collaboration in civil science between scientists of countries of the Euro-Atlantic Partnership Council. The types of scientific meeting generally supported are "Advanced Study Institutes" and "Advanced Research Workshops", and the NATO Science Series collects together the results of these meetings. The meetings are co-organized by scientists from NATO countries and scientists from NATO's Partner countries – countries of the CIS and Central and Eastern Europe.

Advanced Study Institutes are high-level tutorial courses offering in-depth study of latest advances in a field.
Advanced Research Workshops are expert meetings aimed at critical assessment of a field, and identification of directions for future action.

As a consequence of the restructuring of the NATO Science Programme in 1999, the NATO Science Series was re-organized to the four sub-series noted above. Please consult the following web sites for information on previous volumes published in the Series.

http://www.nato.int/science
http://www.springeronline.com
http://www.iospress.nl

Series II: Mathematics, Physics and Chemistry – Vol. 214

Symmetry and Heterogeneity in High Temperature Superconductors

edited by

Antonio Bianconi
University of Rome "La Sapienza",
Department of Physics, Rome, Italy

Springer

Published in cooperation with NATO Public Diplomacy Division

Proceedings of the NATO Advanced Study Research. Workshop on
Symmetry and Heterogeneity in High Temperature Superconductors
Erice, Sicily, Italy
October 4–10, 2003

A C.I.P. Catalogue record for this book is available from the Library of Congress.

ISBN-10 1-4020-3987-5 (HB)
ISBN-13 978-1-4020-3987-4 (HB)
ISBN-10 1-4020-3989-1 (e-book)
ISBN-13 978-1-4020-3989-1 (e-book)

Published by Springer,
P.O. Box 17, 3300 AA Dordrecht, The Netherlands.

www.springeronline.com

Printed on acid-free paper

Printed in the Netherlands.

Contents

Acknowledgements

The Nato Advanced Research Workshop "Symmetry and Heterogeneity in High Temperature Superconductors" held in Erice-Sicily during October 4-10, 2003 has been sponsored by The Nato Science Programme Cooperative Science & Technology Sub-Programme. The workshop has been hosted by the International School of Solid State Physics directed by Giorgio Benedek. We thank the Italian Ministry of Education and University, Sicilian Regional Government Program, and Superstripes-onlus for support. We acknowledge Anna De Grossi for the professional contribution given to the preparation of this book.

Preface

This book is a collection of the papers presented at the workshop on "Symmetry and Heterogeneity in High Tc Superconductors" directed by Antonio Bianconi and Alexander F. Andreev in collaboration with K. Alex Müller and Giorgio Benedek. Philip B. Allen, Neil W. Ashcroft, Alan R. Bishop, J. C. Séamus Davis, Takeshi Egami, Francesco Iachello, David Pines, Shin-ichi Uchida, Subodh R. Shenoy, chaired hot sessione contributing to the success of the workshop.

The object of the workshop was the quantum mechanism that allows the macroscopic quantum coherence of a superconducting condensate to resist to the attacks of high temperature. Solution to this problem of fundamental physics is needed for the design of room temperature superconductors, for controlling the decoherence effects in the quantum computers and for the understanding of a possible role of quantum coherence in living matter that is debated today in quantum biophysics.

The discussions in the informal and friendly atmosphere of Erice was on new experimental data showing that high T_c in doped cuprate perovskites is related with the nanoscale phase separation and the two component scenario, the two-band superconductivity in magnesium diboride and the lower symmetry in the superconducting elements at high pressure.

There has been a large interest in the superconductivity of MgB_2. This system provides the simplest system for testing the high T_c theories, and plays the same role as atomic hydrogen for the development of the quantum mechanics in the twenties. Clear experimental evidence in this system shows that multiband superconductivity enhances the critical temperature from the low T_c range $T_c < 19K$, to the high temperature range, $T_c = 40K$. The heterogeneous structure, the superlattice of superconducting layers, determines the disparity and different spatial location of the Bloch wave functions of electrons at the Fermi level that provides in superconductivity the clean limit. The chemical potential can be tuned by atomic substitutions without increasing inelastic single electron interband scattering. The

Feshbach shape resonance in the exchange-like off-diagonal interband pairing term, as predicted since 1993, appears to be the mechanism for evading temperature decoherence effects and enhancing the critical temperature.

The picture below shows some of the participants at the Erice workshop: 1. Samia Charfi-Kaddour, 2. Cinzia Metallo 3. Laura Simonelli, 4. Sergei Kruchinin, 5. Fedor Kusmartsev, 6. Naurang L. Saini, 7. Alexander Agafonov, 8. Victor Kabanov, 9. Boris Kochelaev, 10. Josef Ashkenazi, 11. Massimo Inguscio, 12. Giorgio Benedek, 13. Francesco Iachello, 14. Karl Alex Muller, 15. Neil W. Ashcroft, 16. David Pines, 17. Antonio Bianconi, 18. Hiroyuki Oyanagi, 19. Nikolay Kristoffel, 20. Takeshi Egami, 21. Sadamichi Maekawa, 22. Anna Maria Cucolo, 23. Kazumi Maki, 24. Georgios Varelogiannis, 25. Shin-ichi Uchida, 26. Annette Bussmann Holder, 27. Roman Micnas, 28. Matteo Filippi, 29. Hidenori Takagi, 30. Fabrizio Bobba, 31. Hugo Keller, 32. John D. Dow, 33. Oystein Fischer, 34. Philip B. Allen, 35. Yuri Latyshev, 37. Sang. W. Cheong, 38. Lev S. Mazov, 39. Christophe Salomon, 40. Nejat Bulut, 41. J.C. Seamus Davis, 42. Toni Schneider, 43. Davor Pavuna, 44. Tomislav Vuletic, 45. Carmine Antonio Perroni, 46. Alan R. Bishop, 47. Efthymios Liarokapis, 48. Subodh R. Shenoy, 49. Pavol Banacky, 50. Jorgen Haase

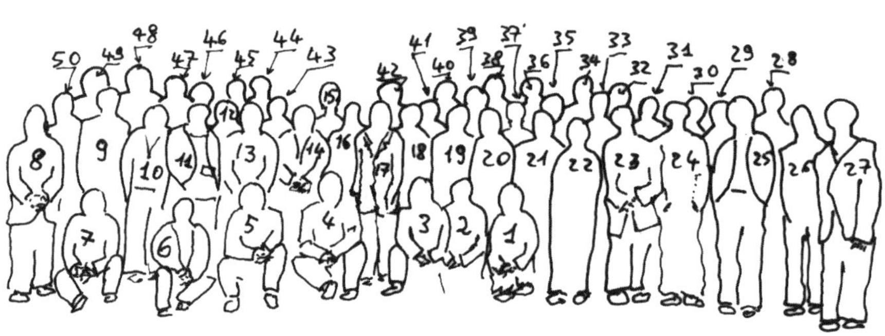

I

ELECTRONIC AND EXCHANGE-LIKE PAIRING SCENARIOS

I.1

SYMMETRY AND HIGHER SUPERCONDUCTIVITY IN THE LOWER ELEMENTS

N. W. Ashcroft
Laboratory of Atomic and Solid State Physics, Cornell University, Ithaca, NY 14853-2501, USA

Abstract: At one atmosphere 29 elements are classified as superconductors; at high pressures there are to date an additional 23, many of these being drawn from the lighter elements. The number of superconductors for the elements in combination appears to be illimitable. Observed symmetries in these systems generally include orderings in both nuclear and electronic degrees of freedom. The fluctuations impelling order in the electron sub-system include those originating with the nuclear degrees of freedom but also with the electrons themselves, both itinerant and localized. For the elements in combination coherent multipole fluctuations in localized states may arise, and the relative contributions of such excitations to electron-pairing is then of some especial interest. When the elements are placed in combination the effects of external pressure may be replicated in part by an equivalent internal pressure, this resulting from a form of chemical pre-compression.

1. INTRODUCTION

Perhaps the most striking early claim of high temperature superconductivity came in 1946 [1] for the light elements *H, N* and a lower alkali metal *M,* all in combination. In experiments involving the metal ammines $M(NH_3)_x$ which remain controversial [2-5], R.A. Ogg [1] invoked the elegant Kamerlingh-Onnes ring geometry for his samples and when these

A. Bianconi (ed.), Symmetry and Heterogeneity in High Temperature Superconductors , 3–20.

were quenched cooled in an external magnetic field a number of them displayed persistent currents even at temperatures as high as 180K. Ogg's quite prescient comment was that "the probable explanation is to be found in trapped electron pairs, recently demonstrated to be stable in fairly dilute [i.e non-metallic] metal ammonia solutions". Under ordinary conditions these are in states of wholly continuous symmetry. This appears to be the first recognition of the importance of electron pairing and its deeper relation (via Bose-Einstein condensation, as Ogg had it) to the extraordinary phenomenon of superconductivity.

Though Ogg's viewpoint has evidently never gained wide acceptance (the matter has been discussed in depth by Edwards [6]), there are certain aspects of his motivating systems (the metal ammines) which possess exceedingly interesting properties. One only has to observe the appearance of a high density of hydrogen implying the presence in a non-diffusive environment of lattice dynamical modes with very significant energies. Further, for *H* in combination with *N,* there is an electronic charge distribution with a scale of linear dimension that implies the possibility of a substantial dipole polarizability. This manifestation of localized charge, and its internal fluctuations, can lead (especially in a crystalline environment) to coherent wave-like excitations to which, as with their phonon counterparts, valence electrons can couple.

From the standpoint of symmetry, and its breaking, the issues raised by these light element systems, and many like them, are several. It will be convenient to introduce and discuss them starting with a pure single-component system, but keeping firmly in mind that for contributions to electron pairing arising from fluctuations in localized electronic charge, it is the groupings of the elements in combination that may well be best placed to achieve this.

2. STARTING HAMILTONIANS AND THEIR SYMMETRY

Beginning therefore with an element (atomic number Z_α), and with one- and two-particle densities $\hat{\rho}_\alpha^{(1)}(\mathbf{r})$ and $\hat{\rho}_\alpha^{(2)}(\mathbf{r},\mathbf{r}'))$ as appropriate observables for macroscopic systems, the Hamiltonian for a neutral ensemble of nuclei $(\alpha = n)$ and electrons $(\alpha = e)$ established in a volume V is, in three-dimensions,

$$\hat{H} = \hat{T}_n + \hat{T}_e + \\ +(1/2)\int_V d\mathbf{r}\int_V d\mathbf{r}'\upsilon_c(\mathbf{r}-\mathbf{r}')\left\{Z_a^2\hat{\rho}_n^{(2)}(\mathbf{r},\mathbf{r}') - 2Z_a\hat{\rho}_n^{(1)}(\mathbf{r})\hat{\rho}_e^{(1)}(\mathbf{r}) + \hat{\rho}_e^{(2)}(\mathbf{r},\mathbf{r}')\right\} \quad (1)$$

where, $\upsilon_c(\mathbf{r}-\mathbf{r}') = e^2/|\mathbf{r}-\mathbf{r}'|$ is the fundamental Coulomb interaction. Here $\hat{T}_n$ and $\hat{T}_e$ are respectively, the kinetic energy operators for N nuclei of mass m_n and $Z_a N$ electrons of mass m_e, i.e in terms of the linear momentum operators $\hat{p}_{\alpha i}$, $\hat{T}_\alpha = \sum \hat{p}_{\alpha i}^2/2m_\alpha$. Since the system fundamentally involves long-range interactions, an alternative way of writing (1) is

$$\hat{H} = \hat{T}_n + \frac{1}{2}\int_V d\mathbf{r}\int_V d\mathbf{r}'\upsilon_c(\mathbf{r}-\mathbf{r}')\left\{Z_a^2\hat{\rho}_n^{(2)}(\mathbf{r},\,\mathbf{r}') - 2Z_a\hat{\rho}_n^{(1)}(\mathbf{r})\bar{\rho} + \bar{\rho}^2\right\} \quad (2a)$$

$$+T_e + \frac{1}{2}\int_V d\mathbf{r}\int_V d\mathbf{r}'\upsilon_c(\mathbf{r}-\mathbf{r}')\left\{\hat{\rho}_e^{(2)}(\mathbf{r},\,\mathbf{r}') - 2\hat{\rho}_e^{(1)}(\mathbf{r})\bar{\rho} + \bar{\rho}^2\right\} \quad (2b)$$

$$-\int_V d\mathbf{r}\int_V d\mathbf{r}'\upsilon_c(\mathbf{r}-\mathbf{r}')\left\{Z_a\hat{\rho}_n^{(1)}(\mathbf{r}) - \bar{\rho}\right\}\left\{\hat{\rho}_e^{(1)}(\mathbf{r}') - \bar{\rho}\right\} \quad (2c)$$

where, apart from endowed statistics, (2a) and (2b) represent formally equivalent quantum mechanical problems each well defined in the thermodynamic limit and differing only according to length scale. In (2) the quantity $\bar{\rho} = NZ_a/V$ (in the limit $NZ_a \to \infty, V \to \infty$, and $NZ_a/V \to \bar{\rho}$). The form of the coupling in (2c) arises when the two sub-systems are taken to occupy a common volume (the backgrounds then cancelling). Starting with (1) (or (2)) the fundamental problems of the physics of condensed matter are to determine ground and excited states and non-equilibrium properties in this limit and, importantly for what follows, the states of order or broken symmetry. It is clear that according to context the orderings may be brought about by changes in average density (i.e in V at fixed N, say) or by changes when the system is in contact with a supplied heat bath (i.e in temperature for a canonical arrangement), or both.

It is immediately evident that in this same limit (1) and (2) possess both continuous translational and rotational symmetry. Yet it is a matter of common experimental experience that at sufficiently low temperature (with the exception of the helium at normal pressures), this symmetry in the nuclear degrees of freedom is broken in the states taken up by (1) or (2), specifically in the profusion among the elements of crystalline phases. There is also no specific reference to the spin of the electrons (or of the nuclei). In the presence of a magnetic field, with associated vector potential $\mathbf{A}(\mathbf{r})$ the modifications to (1) or (2) follow from the minimal substitutions $\hat{p}_{\alpha i} \to \hat{p}_{\alpha i} - (e_\alpha / c)\mathbf{A}$, and the ensuing Hamiltonian continues to preserve some of the original symmetry of H. But once again this symmetry can be broken and (1) admits of phases with spontaneous magnetic order (e.g. for $Z = 24$, under ordinary conditions). This order resides in the electronic spin degrees of freedom, and in a comparative sense this elementary observation now raises the central questions to be addressed here, namely whether further order in the spatial degrees of freedom may arise and then whether order involving both may occur. Independent of the symmetry of the developing phases the form of (1) suggests the emergence of scaling laws for the thermodynamic and correlation functions (and these have been verified explicitly for $Z_a = 1[7]$).

Before proceeding, it may be noted that for $Z_a = 1,(1)$ (or (2)) represents the Hamiltonian for hydrogen, where m_n/m_e takes its lowest value, namely 1836, and where quantum effects of the nuclei are most prominent. The case $Z_a = 3$ corresponds to the first metal in the periodic table under normal conditions, and it will be discussed further below. For $Z_a = 1$ the scale of the associated collective modes is crucial to the possibility of superconducting states of hydrogen when it enters a metallic phase. For values of Z_a larger than unity, and especially for elements taking up metallic states at one atmosphere, (1) leads to a class of electronic states (the 'core states' possessing their own internal fluctuational physics) significantly localized around nuclei, but not especially different in terms of local electronic density from their strictly atomic counterparts. In fact, from the tight binding perspective these states are not expected to change appreciably with currently attainable pressures which presently impel reductions in linear dimensions of a factor of two or more. However (and a key point for what follows), these changes can nevertheless lead to significant outer core-state overlaps, and the consequence of this when coupled with the requirement that the valence electrons states remain orthogonal to the core-states, is of considerable physical importance.

Obviously the electrons most affected in the process of formation of the condensed state are the valence electrons and in the lighter elements the primary physical effects of the remaining (core) electrons is often included through the concept of a pseudo potential, generally non-local. With this understanding Hamiltonian (1) is then modified to reproduce simply the valence electron spectrum. To within density dependent constants (1) is therefore replaced by

$$\hat{H} = \hat{T}_n + \hat{T}_v + \\ + \frac{1}{2}\int_V d\mathbf{r} \int_V d\mathbf{r}' \upsilon_c(\mathbf{r}-\mathbf{r}')\left\{Z_v^2 g(\mathbf{r},\mathbf{r}')\hat{\rho}_n^{(2)}(\mathbf{r},\mathbf{r}') - 2Z_v f(\mathbf{r},\mathbf{r}')\hat{\rho}_n^{(1)}(\mathbf{r})\hat{\rho}_v^{(1)}(\mathbf{r}') + \hat{\rho}_v^2(\mathbf{r},\mathbf{r}')\right\} \quad (3)$$

where Z_v is the long range charge associated with an ion, and $g(\mathbf{r}, \mathbf{r}')$ represents the corrections to pure Coulombic behavior originating with core electrons. In a similar way the term $f(\mathbf{r},\mathbf{r}')$ represents the non-local pseudopotential modifications to point-ion behavior. The familiar band-structure problem emerges from one-electron approximations to (3) when the masses of the nuclei are infinite and the one- and two-particle density operators are the *c*-numbers corresponding to coordinates simply taken as the fixed sites of a perfect crystalline structure.

Though the Hamiltonians are modified (in proceeding from (1) to (3)) the same basic physical questions clearly obtain with respect to the nature of both equilibrium and non-equilibrium states of $\hat{H}$, and of states of order or broken symmetry. Once more the possibility of spontaneous magnetic order is evident (this again residing predominantly in the valence electron structure). Hamiltonians (1), (2), and (3) can immediately be generalized to multielement systems, and the known states of broken electronic symmetry then include spin-density and charge-density wave phases, ionic systems, and so on. These might be referred to as states of diagonal long range order, a familiar classical concept.

3. THE OCCURRENCE OF SUPERCONDUCTING ORDER

The pertinent issue here concerns the possibility of off-diagonal long-range electronic order in the valence electron system, especially in the lighter elements and in their combinations. Of particular importance is the role played by the choice of average density or equivalently the average inter-ion spacing. To approach this problem for the valence electrons, note that underlying (3) is the quite standard quantum mechanical problem obtained by replacing fixed ions by a rigid, continuous and uniform charge distribution, $e\bar{\rho}_v = eNZ_v/V$.

The result, see also (2a), is

$$\hat{H}_v = \hat{T}_v + \frac{1}{2}\int_V d\mathbf{r}\int_V d\mathbf{r}'\upsilon_c(\mathbf{r}-\mathbf{r}')\left\{\hat{\rho}_v^{(2)}(\mathbf{r},\mathbf{r}') - 2\hat{\rho}_v^{(1)}(\mathbf{r})\bar{\rho}_v + \bar{\rho}_v^2\right\} \quad (4)$$

the much studied interacting electron gas problem. The constants of this Hamiltonian $(\hbar, m_e, e)$ define the familiar atomic unit of length $a_0 = \hbar^2/m_e e^2$, in terms of which the standard linear measure r_s of average inverse valence electron density (through $(4\pi/3)r_s^3 a_0^3 = 1/\bar{\rho}_v$) emerges. Though formally independent of spin (4) admits of magnetically ordered states for sufficiently dilute conditions (or large r_s). It also takes up prominent states of continuous symmetry (the Fermi liquid for $r_s \sim O(1)$) and, again for sufficiently large r_s, states of broken translational symmetry (the Wigner crystal, or a Wigner crystal with a basis). The central issue for the viewpoint to follow is whether states of (4) can be found that spontaneously break a gauge symmetry in the presence of a magnetic field. As will be seen this may well be the case and the question can therefore be posed again, first for Hamiltonian (3) representing the particular static periodic system for which $\langle\bar{\rho}_e^{(1)}(\boldsymbol{r})\rangle$ conforms to this symmetry, and then finally for (1). According to chosen conditions the latter can again represent a periodic system but only on time scales long compared with mean phonon periods. As noted above a control parameter is implicit in this analysis and it is mean density, this presently being an experimental variable of some consequence.

An *in principio* answer to the fundamental question on the symmetry of the ground state of the interacting electron gas was provided in 1965 by Kohn and Luttinger [8] who argued that pairing ground states for the homogeneous interacting electron gas might well be preferred over the normal Fermi liquid (but in energetic measures not greatly). In terms of static interactions their argument was centered in part on the presence of

Friedel oscillations (and hence attractive regions) in effective electron-electron interactions. More formally, the kernel of the Eliashberg equation contains contributions (especially from ladder diagrams) that can very much favor formation of a paired state [9-11]. In fact this may even be seen as the second of two possible symmetry breakings actually 'detected' by the presence of a vector potential, **A**. The first involves its role in detecting an insulator to metal transition. Thus when (4) is augmented to include the effects of **A**, it can be shown [12] that the consequent ground-state energy per electron (i.e $< H_\upsilon > /N$) satisfies

$$\left(\frac{d}{dr_s}\right)\left(\frac{<\hat{H}_v>}{N}\right)+\frac{1}{r_s}\left(\frac{<\hat{T}_v>}{N}+\frac{<\hat{H}_v>}{N}\right)=\frac{1}{r_s}\left(\frac{V}{N}\right)\frac{\mathbf{A}\cdot\mathbf{J}}{c} \tag{5}$$

where **J** is the current density in the presence of **A**. Precisely the same result obtains [12] for a charged two-component system (corresponding, for example, to (1)); the result is exact and, importantly, it hold independent of the symmetry of the states actually taken up, in particular for states displaying off-diagonal long-range-order. But among such phases might well be in insulating states, in which case **J** = 0, necessarily. If so then so far as the energy is concerned, and for the limiting case of a system with the Kamerlingh-Onnes ring topology mentioned earlier, an insulating state will then not detect the presence of **A**, an important distinction first emphasized by Kohn [13]. As has been emphasized in Ref [12], a transition from insulating to metallic state can actually be viewed as the breaking of a global gauge symmetry (and the condensation of gauge bosons). Accordingly given the breaking of a gauge symmetry associated with formation of the superconducting state in the presence of a magnetic field, it may be interesting to seek a deeper connection between superconductivity and the metal-insulator transition [12].

This notion evidently gains more prominence when the same question is asked of Hamiltonian (3); in a one-electron approximation this eventually leads, as noted, to band-structure. For the light elements, or those dominated by *s-p* character, treatments of (3) involving structural perturbation theory have been generally successful in accounting for observed atomic arrangements. Nevertheless, the possibility of inherent electronic instability was recognized, particularly by Overhauser [14], and notably that the states could be susceptible to exchange driven electronic transitions leading either to spin-density waves or charge-density waves. Utilizing a Hubbard approach, at fixed static structure, Siringo et al [15] observed that commensurate charge- density-waves might even develop in the alkali series leading to a metal-insulator transition at high compression. But when full relaxation of nuclear coordinates is permitted, very significant structural complexity seems to occur, at least for infinitely massive nuclei. This has been predicted to be the case for lithium and sodium [16, 17], and is even observed to be the case in the higher s-p alkali series.

Yet the primary issue still remains, namely whether full restoration of electron-electron interactions can again lead to pairing ground-states for the valence electrons but now under far more general conditions, and especially whether periodicity in the underlying lattice, and the possible occurrence of complex structures just mentioned, can enhance this prospect. Kohn and Luttinger's question in a band context has quite interesting features since the effective interaction can be appreciably modified in a system with discrete translational symmetry [18]. This is notably so in multi-band systems [19-23] and especially the case when electron and hole bands are both prominent in the single-particle electronic structure. Indeed, the most effective situation appears to be the case where the system is compensated; here fluctuations are necessarily of a compensated correlated form, and these actually lead to attractive contributions to the effective electron-electron interaction [24]. When applied to the case of a proton-paired metallic phase of hydrogen, the enhancements to the predicted transition temperatures are significant [25].

Thus, the simplest and earliest approach to the inclusion of many-electron effects is the Thomas-Fermi (TF) method, and it is immediately useful in the scale it sets when electron-electron interactions are restored. In a BCS viewpoint the measure of direct electron-electron repulsion is

$$\mu = \langle N_0 v_c(\mathbf{k}'-\mathbf{k}) \rangle \tag{6}$$

the average being taken over a spherical Fermi surface whose diameter is $2k_F$ In the above $N_0 = (1/4\pi^2)k_F^3/\varepsilon_F$ is the (intensive) density of states per unit volume, for a given spin, evaluated at the Fermi energy ε_F. When many-body effects are treated in the TF approximation, and wavevectors are normalized to $2k_F(\mathbf{x} = \mathbf{k}/2k_F)$, and $(k_0^2/2k_F) = \gamma = (4/9\pi^4)^{1/3} r_s$, where k_0 is the Thomas- Fermi wave-vector), then

$$N_0 v_c(\mathbf{x}-\mathbf{x}') = (\gamma^2/2)\big/\left((\mathbf{x}'-\mathbf{x})^2+\gamma^2\right) \tag{7}$$

the average of this over a sphere of unit diameter then giving $(\gamma^2/2)ln(1+1/\gamma^2)$. Inclusion of retardation effects then leads to a first estimate for the Coulomb pseudopotential namely;

$$(1/\mu^*) = \left[(\gamma^2/2)ln(1+1/\gamma^2)\right]^{-1} + ln(\hbar\omega_{p,e}/\hbar\omega_p) \tag{8}$$

where $\hbar\omega_{p,e}$ is the electron plasmon energy and $\hbar\omega_p$ is the corresponding quantity for the ions. Standard estimates for these are already sufficient to yield the well known values $\mu^* \sim 0.1$; for μ itself the figure is ~ 0.2.

Detailed inclusion of electron dynamics (well beyond TF) can lead to further reductions in μ^*. Intrinsic pairing (i.e where the effective electron-electron interaction acquires no additional enhancements from, for example, phonon based fluctuations) has been studied in some detail for single-band systems as noted [9-11], but only modestly for multi-band situations. The pairing tendencies also seem very much enhanced when dimensionality is reduced [25,26]. Included under this broad rubric would be the electronic fluctuation that arises from the charge corresponding to the class of electron states classified above as localized and specifically associated with bound states. Though localized plasmons might exist in principle for such states (these attributable to dynamic monopoles) a further important class of excitations (they are also propagating) is the set of polarization waves associated with localized charge but periodically arranged [28] (these attributable to dynamic multipoles). The energies of these are formally contained within appropriate ground state functionals, but it is known now that their development requires a non-local treatment of such functionals (Van der Waals attraction, the most prominent manifestation of such effects, mandates correlated dipolar fluctuations). For the present it may be noted that quantized waves of polarization are coherent and are clearly synthesized from such correlated fluctuations.

4. NUCLEAR AND ELECTRONIC FLUCTUATION COMBINED

These correlated fluctuations themselves 'ride' on a further set of coherent fluctuations taking place at a much lower frequency scale and normally attributed to the phonons, the traditional exchange Bosons associated with superconductivity. Real systems are never devoid of ionic or nuclear motion, and at the very least it is now Hamiltonian (3) (and eventually its extension to alloys) that applies for a full discussion of superconductivity; density fluctuations in the nuclear coordinates are omnipresent and of course their effects on electronic ordering have been evident for quite some time. An elementary estimate of the relative importance of (monopole) polarization arising from phonons and the (multipole) equivalents arising from internal fluctuations, primarily of a dipole character, can now be easily given.

First, valence and core electrons are formally identical; however, the separation of valence and core electron density is dictated by the standard view of atomic physics. Thus for an ion at j, coordinates $\mathbf{r}_{j,i}$ are assigned to the $Z_a - Z_v$ electrons designated as core electrons, the understanding being that the states of the system are such that $<\mathbf{r}_{j,i}^2>^{1/2}$ is a small quantity ($\lesssim a_0$). Thus if $\mathbf{u}_j$ *is* the displacement of an ion at site j arising from phonons, then the one-electron density operator is approximated by

$$\hat{\rho}_e^{(1)}(\mathbf{r}) = \sum_i \delta(\mathbf{r} - \mathbf{r}_{e,i}) + \sum_j \sum_i \delta\left(\mathbf{r} - \mathbf{R}_j - \mathbf{u}_j + \hat{d}_j / e\right) \quad (9)$$

where the second term is taken to account for electronic density that will reside in localized states, and the first corresponds to the valence electron density. In the above $\mathbf{R}_j$ is a lattice site, and the dipole operator $\hat{d}_j$ for site j is written as

$$\hat{d}_j = \sum (-e)\mathbf{r}_{j,i} \quad (10)$$

It is useful to re-emphasize at this point that the generalized Kohn-Luttinger question is being asked for the case where all the $\mathbf{R}_j$ are rigorously zero (i.e for the possibility of intrinsic pairing in the valence electrons arising from fluctuation in both itinerant and localized charge).

But next, consider the characteristic scales of $\mathbf{u}_j$ and $\hat{d}_j$; these follow from noting that

$$\langle u_j^2 \rangle^{1/2} \sim (m_e/m_n)^{1/2} \left(\frac{e^2/2a_0}{\hbar\omega_D} \right)^{1/2} a_0 \qquad (11)$$

where $\hbar\omega_D$ is a typical phonon energy, whereas

$$\langle \hat{d}_j^2 \rangle^{1/2} / e \sim \left(\frac{\hbar\Omega}{e^2/\alpha^{1/3}} \right)^{1/2} \left(\frac{\alpha}{a_0^3} \right)^{1/3} a_0 \qquad (12)$$

where $\hbar\Omega$ is a typical excitation of an ion (or ion complex) whose polarizabity is α. Though the time scales of the two classes of excitations differ appreciably (as they do between electrons and phonons) it should be noted, once again, that both are coherent in a crystalline environment. The main point, however, is that typical excitation energies $\hbar\Omega$ may be an notable fraction of an atomic unit, and for ions with significant core spaces the dipole polarizability can also reach appreciable fractions of a_0^3. Thus depending on system $\langle \mathbf{u}_j^{\,2} \rangle^{1/2}$, and $\langle \mathbf{d}_j^{\,2} \rangle^{1/2} / e$ can be comparable and it is apparent that interference between these terms will not always be constructive (this could well be the case in a simplified view of the noble metals, for example, where a tight-binding view of the *d*-electrons is taken).

With the scale of lattice displacement approximately established it is also useful to recall that phonons have been the traditional and indeed dominant mechanism for pairing, and a supporting argument for this is usually to be found in the normal state transport properties, for example the static resistivity. Thus in the static resistivity common arguments hold that at low temperatures and in three-dimensions the number of phonons available to scatter electrons rises as T^3. Of these a fraction $\sim T^2$ will satisfy the constraint restricting scattering to a (Fermi) surface. For normal scattering processes, the quasi-classical Boltzmann equation introduces a factor $(1 - cos\Theta) \sim q^2 \sim T^2$ into the determination of the actual current density. Here Θ is the angle between velocity vectors before and after scattering. Finally, the electron-phonon interaction leads to a scattering rate $\sim q \sim T$ and hence for normal processes to an overall T^5 rise in resistivity (the Bloch-Grueneisen 'law'). Note, however, that for strong interband or Umklapp processes, and a Fermi surface in many sheets (and possibly with varying effective masses) the factor $(1 - cos\Theta) = 2sin^2\Theta/2$ will lead instead to averages close to a constant (the velocities which are normal to the constant energy surface now suffer large relative changes in the scattering event).

This last argument may be of particular relevance to a layered or near two-dimensional system, for here the Fermi surface can become a sequence of Fermi cylinders, and of the now order T^2 phonons only a measure $\sim T$ will satisfy the constraints to lie on these cylinders. And if the degree of doping is such that the Fermi curves again lie in several zones, and if again the scattering is strong, the factor $(1 - cos\Theta)$ for transitions between different bands will once more average to an approximate constant. Finally, for scattering that is being dominated by electron-phonon interactions favoring such interband effects, the corresponding scattering rate is once more proportional to K (a reciprocal lattice vector). Overall in near two-dimensional system, the resistivity should therefore be roughly proportional to T (also the high temperature limit) but with minor corrections anticipated to account for the expected departure from absolute Fermi curves in what is a strictly three-dimensional environment, these clearly being dependent on the degree of doping. When conditions are such that normal process do eventually dominate, it is clear that a higher dependence on T will again be expected (possibly T^4), but over a relatively small range of temperatures.

It is evident that if intrinsic electronic effects are insufficient to bring about a superconducting instability, then additional contributions to pairing may be sought in these phonon terms whose presence should be revealed in normal state transport, as described above. Traditionally the measure of possible phonon attractive contributions, $< N_o V_{ph} >= \lambda$, originating with screened electron-phonon coupling $g_{\mathbf{q}}(\mathbf{k}', \mathbf{k})$ associated with the scattering of an electron from $\mathbf{k}$ to $\mathbf{k}'$, by a phonon of wave-vector $\mathbf{q}$. In this case the average is required of

$$-N_0 2|g_{\mathbf{q}}(\mathbf{k}', \mathbf{k})|^2/\hbar\omega\ (\mathbf{k}'-\mathbf{k}) \qquad (13)$$

where for a Debye spectrum and longitudinal modes for nuclei of mass Am_n

$$g_{\mathbf{q}}(\mathrm{k}',\mathrm{k}) = \frac{4\pi e^2 f(\mathbf{k}',\mathbf{k})}{\left((\mathbf{k}'\text{-}\mathbf{k})^2 + k_0^2\right)} i(\mathbf{k}' - \mathbf{k}) \cdot \left(\frac{\hbar}{2cq} \cdot \frac{N/V}{Am_n}\right)^{1/2} \tag{14}$$

Here the actual sound speed will be written as $c = \alpha\, v_F \left\{3m_e/Z\, Am_m\right\}^{1/2}$ the role of the dimensionless α being to correct the standard Bohm-Staver estimate. For *normal* intraband ($\mathbf{k}'-\mathbf{k} = \mathbf{q}$) processes, the contribution to $< N_0 V_{ph} >$ follows from an average on a sphere of unit diameter (but with the restriction $q = |\mathbf{k}' - \mathbf{k}| < k_D$, also normalized) of

$$-N_o \upsilon_c (\mathbf{x} - \mathbf{x}') \left\{ \frac{f^2(\mathbf{x}',\mathbf{x})}{\alpha^2} \cdot \frac{\gamma^2}{(\mathbf{x}' - \mathbf{x})^2 + \gamma^2} \right\} \tag{15}$$

The quantity α is < 1; for elevated densities f increasingly reflects the short range repulsive region of the pseudopotential and it can become appreciable. As is well known, depending on system the phonon mechanism may well prevail over the direct electron repulsion term but that inclusion of Umklapp terms ($\mathbf{k}' - \mathbf{k} = \mathbf{q} + \mathbf{K}$, with $\mathbf{K}$ a reciprocal lattice vector) can increase the likelihood considerably. From the definition of N_o, (and for NZ electrons in a volume V) it follows that $|g_{\mathbf{q}}(\mathbf{k}', \mathbf{k})|^2 \sim (V/ZN)(\varepsilon_F.\,\hbar\omega_D)$ and g is therefore proportional to the geometric mean of the electron and phonon energy scales. The coupling can therefore be large and, again depending on system, the phonon-term $\lambda = (N_o V_{ph})$ may then approach the strong-coupling values ~ 1.

For a full discussion of the theory underlying determinations of T_c from phonon based mechanisms, particularly within the Eliashberg framework, the reader is directed to the review by Allen and Mitrovic [29]. Here the issue devolves on the essential input, arising from the interactions defining the problem at hand (and subsequently entering the kernel of the linearized Eliashberg equations) and on the role of average density with its ability to alter the relative contributions of valence electron coupling to the various excitations discussed above. In the quest for off-diagonal-long-range-order for the valence electrons (and hence in formulating the establishment of the overall effective electron-electron interaction) the sources to be examined are (i) the valence electrons themselves, but in a periodic arrangement and in multiple bands, (ii) the internal dynamics of the core states and their coherent excitations emerging once more from a time average periodic environment, and (iii) the phonons which are traditionally treated within an approximation of rigid, or near rigid ions. It is to be emphasized again that in atomic complexes, with sizable spatial scales, the contributions from (ii) may be especially significant. The time scales of these excitations are quite disparate and this is already encountered in the comparison of electron plasmon frequencies and typical Debye frequencies. Likewise there is a significant difference between the times scales of phonons and polarization waves, but not especially between polarization waves and plasmons. It should be noted that all the excitations suffer damping. And it should also be particularly noted that spin fluctuations in the valence electron system can also be important to the pairing problem.

5. PRESSURE, ELECTRONIC STRUCTURE, AND OFF-DIAGONAL-LONG-RANGE-ORDER

On the basis of an assessment of the rankings in energy of the various terms entering (1), or its reductions in a periodic system, it has been suggested [18] that the problem of superconductivity could be approached via Eliashberg theory by starting first with such fluctuational attraction as may arise from *all* electrons in a crystalline space, and only later augmenting these with the additional interactions arising from phonons and, as indicated above, from internal fluctuations of localized electronic charge. The simplest cases to consider are systems where the latter can be neglected (generally where $\hbar\Omega$ is large but α is compensatingly small). This viewpoint can change considerably however when the system is constructed from such elements in combination.
The case $Z = 3$, lithium, conforms well to this approximation, for which (1) reads

$$\hat{H} = \hat{T}_n + \hat{T}_\upsilon + (1/2)\int_V d\mathbf{r}\int_V d\mathbf{r}'\,\upsilon_c(\mathbf{r}-\mathbf{r}')\{9\hat{\rho}_n^{(2)}(\mathbf{r},\mathbf{r}') - 6\hat{\rho}_n^{(1)}(\mathbf{r})\hat{\rho}_e^{(1)}(\mathbf{r}) + \hat{\rho}_e^{(2)}(\mathbf{r},\mathbf{r}')\} \quad (16)$$

and (3) represents an effectively monovalent system described by

$$\hat{H} = \hat{T}_n + \hat{T}_\upsilon + \\ +(1/2)\int_V d\mathbf{r}\int_V d\mathbf{r}'\,\upsilon_c(\mathbf{r},\mathbf{r}')\{g(\mathbf{r},\mathbf{r}')\hat{\rho}_n^{(2)}(\mathbf{r},\mathbf{r}') - f(\mathbf{r},\mathbf{r}')\hat{\rho}_n^{(1)}(\mathbf{r}')\hat{\rho}_v^{(1)}(\mathbf{r}) + \hat{\rho}_v^{(2)}(\mathbf{r},\mathbf{r}')\} \quad (17)$$

In the supposed limit $m \to \infty$ and for crystalline symmetry, the one-electron approximation to (17) leads at standard densities to the familiar band structure description of metallic lithium, originally studied at high density by Boettger and Trickey [29]. Recent total energy treatments [16] of (17) (via density functional methods) show that at one atmosphere the ground-state structure taken up conforms to 9R ($\delta - Sm$); at progressively higher pressures fcc is first preferred, in agreement with experiment. However, at around 40 GPa several quite complex structures are predicted to become competitive, and indeed complex structures are found experimentally [30]. Most interestingly for the subsequent problem of broken electronic symmetry, the valence bands are *diminishing* in width at the corresponding densities, an effect attributable to valence-core orthogonalization which progressively diverts valence electron density into interstitial regions.

Elementary arguments now strongly suggest [16] the possible onset of off-diagonal-long-range-order in the valence electrons once $\hat{T}_n$ is restored, high dynamical energies then being expected at elevated densities. First, as a consequence of band narrowing the density of states is expected to increase. Second, steady reduction in inter-ion separation carries with it the expectation that the repulsive core region of the pseudopotential will gain increasing prominence; in other words the factor f in (17) will lead to a generally stronger electron-ion interaction and hence a stronger electron-phonon interaction. This is not unlike the situation encountered in dense hydrogen, which leads to the prediction of elevated superconducting transition temperatures for metallic states. Here the same arguments prevail, except that should lithium take up a complex structure with an *even* number of ions per cell, then the ensuing complex compensated metallic system further favors pairing through the intrinsic mechanisms discussed above. Experimentally, lithium now seems to be an element with one of the highest of transition temperatures, this first being measured (with a resistive technique) by Shimizu et al [31], and later with an inductive technique by Struzhkin et al [32], and later still with an inductive method in a near hydrostatic environment by Deemyad and Schilling [33]. The measurements show a clear progression of phases clearly illustrating the role of pressure in tuning both structure and the contributions to pairing from nuclear and electronic sources. But Deemyad and Schilling [33] also report that above 67 GPa the observation of superconductivity abruptly disappears, and if substantiated this raises the prospect of further changes in symmetry, including even the possibility of a transition from a metal to an insulator upon *increase* of density [16].

It is evident that the arguments just outlined for lithium (these being invoked even earlier for hydrogen) can also be invoked for certain light elements in combination. The most interesting class appears to be those hydrogen rich compounds that are actually dominated by hydrogen, these eventually being driven metallic by steady increase in density. Attention is then focussed on the Group IVa hydrides which, among chemically simple systems are among the most hydrogen dominant as may be found [22]. Further, upon an assumption of overlapping bands, they will contain 8 electrons per cell in simple structures, exactly as is the case for MgB_2 at one atmosphere. What endows these systems with especial physical interest is the fact that prior densification of hydrogen is being attained through the presence of other constituents; it is a form of chemical pre-compression. As a commentary on the likely role of the lower modes in these alloys it may be noted that tin and lead as pure constituents are among the strong coupling superconductors, and silicon and germanium also have notable superconducting transition temperatures when, impelled by pressure, they enter the metallic state.

The Group IVa hydrides include methane, silane, germane, stannane and plumbane, the last being marginally stable but nevertheless an important candidate for study at high pressures. These encompass a considerable spread in mass and hence through systematic deuterations they offer, in metallic states, a possibly unique opportunity to assess broken gauge symmetry and the theory of strong coupling superconductivity in general. An equally important point is that the pressure required to attain a metallic state with r_s typically around 1.5 is considerably less than is required for pure hydrogen itself [34] providing only that there is no disproportionation. Given the fact that an even number of electrons will be found in unit cells, the metals will be compensated and again this presents a situation especially favorable to a reduction in Coulomb pseudopotentials.

Arguments for considering *ternary* Group IVa hydride alloys can also be advanced [34] in order to maximize the contributions of lower modes. An interesting question then centers on symmetry; will such arrangements favor stoichiometric arrangements or, reverting to the initiating example (the quenched ammines) will disorder continue to play an as yet unexplained microscopic role in superconductivity? From the example of the Group IVa hydrides given above it would appear that if the relatively small gaps separating the Li 2s bands and the hydrogen bands (as reported by Kohanoff *et al.* [35]) could be closed by application of pressure, then the common bands should begin to exhibit a degree of similarity to the hydrides. In particular the hydrogens are now in an environment of itinerant electrons. Accordingly for $Li(NH_3)_4$ and indeed for many of the metal ammines, a case can be made for the presence of high temperature superconductivity at higher densities. Note that the remnant polarizability associated with the molecular order can still be expected to be large.

Acknowledgements: This work was supported by the US National Science Science Foundation.

REFERENCES

1. R.A. Ogg, Phys. Rev. 69, 243 (1946).
2. H.A. Boorse, D.B. Cook, R.B. Pontius, and , M.W. Zemansky, Phys. Rev. 70, 92 (1946).
3. J.W. Hodgins, Phys. Rev. 70, 568 (1946).
4. R.A. Ogg, Phys. Rev. 70, 93 (1946).
5. R.B. Gibney, and G.L Pearson, Phys. Rev. 72, 76 (1947).
6. P.P. Edwards, Journal of Superconductivity: Incorporating Novel Magnetism 13, 933 (2000).
7. K. Moulopoulos, and N.W. Ashcroft, Phys. Rev. B 41, 6500 (1990).
8. W. Kohn, and J. Luttinger, Phys. Rev. Lett. 15, 524 (1965).
9. Y. Takada, J. Phys. Soc. Jpn. 45, 786(1978); 61, 238(1992); 61, 3849 (1992).
10. M. Grabowski, and L.J. Sham, Phys. Rev. B 29, 6132 (1984).
11. H. Rietschel, and L.J. Sham, Phys. Rev. B 28, 5100 (1983).
12. K. Moulopoulos, and N.W. Ashcroft, Phys. Rev. B 45, 11518 (1992).
13. W. Kohn, Phys. Rev. 133A, 171 (1964).
14. A.W. Overhauser in Highlights in Condensed Matter Theory, F. Bassani, F. Fumi and M.P. Tosi (eds.) (North Holland, Amsterdam) (1985).
15. F. Siringo, R.Pucci, , and G.N.N. Angilella, High Pressure Res. 15, 255 (1997).
16. J.B. Neaton, and N.W. Ashcroft, Nature 400, 141 (1999).
17. J.B.Neaton, and N.W. Ashcroft, Phys. Rev. Lett. 86, 2830 (2001).
18. C.F. Richardson, and N.W. Ashcroft, Phys. Rev. B 55, 15130 (1997).
19. W.A. Little, Phys. Rev. A 134, 1416 (1964).
20. B.T. Geilikman, JETP 21, 796 (1965).
21. G. Vignale, and K.S. Singwi, Phys. Rev. B 31 2729 (1985).
22. B.K. Chakraverty, Phys. Rev. B 48, 4047 (1993).
23. C.F. Richardson, and N.W. Ashcroft, Phys. Rev. B 54, R764 (1996).
24. C.F. Richardson, , and N.W. Ashcroft, Phys. Rev. Lett. 78, 118 (1997).
25. I.G. Khalil, M. Teter, and N.W. Ashcroft, Phys. Rev. B 65, 195309 (2002).
26. V.M. Galitski, and S. Das Sarma, Phys. Rev. B 67, 144520 (2003).

27. G.S. Atwal, and N.W. Ashcroft, to be published.
28. P.B. Allen, , and B. Mitrovic, Solid State Physics 37, F. Seitz, D. Turnbull, and H. Ehrenreich (eds.) pl. (1982).
29. J.C. Boettger, and S. Trickey, Phys. Rev. B 32, 3391 (1985).
30. M. Hanfland, K. Syassen, N.E. Christensen, and D.L. Novikov, Nature 408, 174 (2000).
31. K. Shimizu, H. Ishikawa, D. Takao, T. Yagi, and K. Amaya, Nature 419, 597 (2002).
32. V.V. Struzhkin, M.I. Eremets, W. Gan, H-K. Mao, and R.J. Hemley, Science 298, 1213 (2002).
33. S. Deemyad, and J.S. Schilling, Phys. Rev. Lett. 91, 167001 (2003).
34. N.W. Ashcroft, Phys. Rev. Lett. 92, 187002 (2004).
35. J. Kohanoff, F. Buda, M. Parrinello, and M.L. Klein, Phys. Rev. Lett. 73, 2599 (1994).

I.2

FESHBACH SHAPE RESONANCES IN MULTIBAND HIGH T_C SUPERCONDUCTORS

A. Bianconi and M. Filippi
Dipartimento di Fisica, Università di Roma "La Sapienza", P. Aldo Moro 2, 00185 Roma, Italy

Abstract: We describe particular nanoarchitectures (superlattices of superconducting wires and layers) where a mechanism to evade temperature decoherence effects in a quantum condensate is switched on by tuning the charge density. The superlattice structure determines the subbands and the corresponding Bloch wavefunctions of charge carriers at the Fermi level *with different parity* and *different spatial locations*. The disparity and negligible overlap between electron wave-functions in different subbands suppress the single electron interband impurity scattering rate and allow the multiband superconductivity in the clean limit. The quantum trick that bestows to the system the property to resist to the decoherence attacks of high temperature is the Feshbach shape resonance in the interband off-diagonal exchange-like pairing term i.e., in the quantum configuration interaction between pairing channels in different subbands. It occurs by tuning the chemical potential at a particular point near a Van Hove singularity (vHs) in the electronic energy spectrum, or a Lifshitz electronic topological transition (ETT), associated with the change of the dimensionality of the Fermi surface topology of one of the subbands.

Key words: Feshbach resonance; Shape resonance; Diborides; Heterostructure at atomic limit.

A. Bianconi (ed.), Symmetry and Heterogeneity in High Temperature Superconductors, 21–53.

1. INTRODUCTION

Understanding the mechanism that allows a quantum condensate to resist to the decoherence attacks of temperature is relevant in many fields going from high T_c superconductivity [1] to quantum computing and quantum biophysics [2]. It is possible that the evolution has refined the living matter to acquire a quantum phase coherence for its biochemical reactions. In fact the living matter displays a dynamical and spatial order with collective properties and non-local interactions as a superfluid. So the research is focusing on the nanoarchitectures deploying quantum tricks for the decoherence-evading qualities at room temperature

The macroscopic quantum phase coherence in superfluids [3-5] appears only below a critical temperature T_c much lower than the earth temperature because the phase coherence is suppressed by decoherence effects at high temperature. While the research for high T_c superconductors has been focused for many years to systems at the borderline between superconducting and ferromagnetic or antiferromagnetic phases it is now shifting toward the identification of a particular nanoscale heterogeneity suppressing decoherence effects at hightemperature. The key for solving the problem could be in subtle structural and electronic details.

Looking for details in the theory of superconductivity one has to go beyond the standard BCS approximations [3] for a generic homogeneous system (i) the high Fermi energy: the Fermi energy is assumed at an infinite distance from the top or the bottom of the conduction band, (ii) the isotropic approximation: the pairing mechanism is not electronic state dependent. The BCS wave-function of the superconducting ground state has been constructed by configuration interaction of all electron pairs (+k with spin up, and -k with spin down) on the Fermi surface in an energy window that is the energy cut off of the interaction,

$$|\Psi_{BCS}\rangle = \prod_k (u_k + v_k c^+_{k\uparrow} c^+_{-k\downarrow})|0\rangle \tag{1}$$

where $|0\rangle$ is the vacuum state, and $c^+_{k\uparrow}$ is the creation operator for an electron with momentum k and spin up. The Schrieffer idea [3] of this state with off-

diagonal long range order came from the configuration interaction theory by Tomonaga involving a pion condensate around the nucleus [6].
In anisotropic superconductivity one has to consider configuration interaction between pairs, in an energy window ΔE around the Fermi level, in different locations of the k-space with a different pairing strength, that gives a k-space dependent superfluid order parameter i.e., a k-dependent superconducting gap.
A particular case of anisotropic superconductivity is multiband superconductivity, where the order parameter and the excitation gap are mainly different in different bands.

The advances in this field are related with the development of the theory of configuration interaction between different excitation channels in nuclear physics including quantum superposition of states corresponding to *different spatial locations* for interpretation of resonances in nuclear scattering cross-section [7] related with the Fano configuration interaction theory for autoionization processes in atomic physics [8].

The theory of two band superconductivity, including the configuration interaction of pairs of oppositive spin and momentum in the *a*-band and *b*-band, was developed on basis of the Bogolyubov transformations [9-13] where the many body wave function [14] is given by

$$\left|\Psi_{Kondo}\right\rangle = \prod_{k}(u_k + v_k a^{+}_{k\uparrow} a^{+}_{-k\downarrow}) \prod_{k'}(x_{k'} + y_{k'} b^{+}_{k'\uparrow} b^{+}_{-k'\downarrow})\left|0\right\rangle \tag{2}$$

The element corresponding to the transfer of a pair from the *a*-band to the *b*-band or vice versa appears with the negative sign in the expression of the energy. This gain of energy is the origin of the increase of the transition temperature driven by interband pairing.

The two band superconductivity has been proposed for metallic elements and alloys [10-42], for doped cuprate perovskites [43-114], for magnesium diboride [115-184] and for few other materials as Nb doped $SrTiO_3$ [185], Sr_2RuO_4 [186-188] YNi_2B_2C, $LuNi_2B_2C$ [189] and $NbSe_2$ [190]. We provide a nearly complete reference list on this subject since it is not available elsewhere.

The multiband superconductivity shows up only in the "clean limit", where the single electron mean free path for the interband impurity scattering satisfies the condition $l > hv_F/\Delta_{av}$ where v_F is the Fermi velocity and Δ_{av} is the average superconducting gap [24,28,30,35].

The criterium that the mean free path should be larger than the superconducting coherence length must be met. This is a very strict condition that implies also that the impurity interband scattering rate γ_{ab} should be very small $\gamma_{ab} << (1/2)(K_B/\hbar)T_c$. Therefore most of the metals are in the "dirty limit" where the interband impurity scattering mixes the electron wave functions of electrons on different spots on bare Fermi surfaces and it reduces the system to an effective single Fermi surface.

The "interchannel pairing" or "interband pairing" that transfers a pair from the "a"-band to the "b"-band and viceversa in the multiband superconductivity theory is expressed by

$$\sum_{k,k'} J(k,k')(a^{+}_{k\uparrow}a^{+}_{-k\downarrow}b_{-k'\downarrow}b_{k'\uparrow}) \qquad (3)$$

where a^{+} and b^{+} are creation operators of electrons in the "a" and "b" band respectively and $J(k,k')$ is an exchange-like integral. This interband pairing interaction may be repulsive as it was first noticed by Kondo [14]. Therefore it is a non-BCS pairing process since in the BCS theory an attractive interaction is required for the formation of Cooper pairs.

Another charateristic feature of multiband superconductivity is that the order parameter shows the sign reversal in the case of a repulsive interband pairing interaction [26].

The non-BCS nature of the interband pairing process is indicated also by the fact that, when it is dominant, the isotope effect vanishes even if the intra-band attractive interaction in each band is due to the electron-phonon coupling.

Moreover the effective repulsive Coulomb pseudopotential in the Migdal-Eliashberg theory is expected to decrease (so the effective coupling strength increases) where the interband pairing is dominant.

In this work we will discuss the particular case of multiband superconductivity where a Van Hove-Lifshitz feature [191] in the electronic energy spectrum associated with a change of Fermi surface topology shown in Fig. 1 occurs within an energy window around the Fermi energy. The width of this windows is the cut-off energy for the pairing that is determined by the energy of the exchanged excitation of phononic or electronic or magnetic origin.

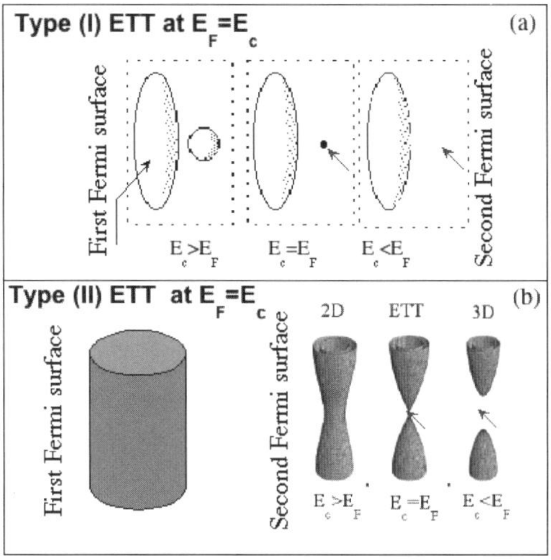

Figure: I:2:1. The different types of 2.5 Lifshitz electronic topological transition (ETT): The upper panel shows the type (I) ETT where the chemical potential E_F is tuned to a Van Hove singularity (vHs) at the bottom (or at the top) of a second band with the appearance (or disappearance) of a new detached Fermi surface region. The lower panel shows the type (II) ETT with the disruption (or formation) of a "neck" in a second Fermi surface where the chemical potential E_F is tuned at a vHs associated with the gradual transformation of the second Fermi surface from a two-dimensional (2D) cylinder to a closed surface with three dimensional (3D) topology characteristics of a superlattice of metallic layers

2. FESHBACH SHAPE RESONANCES

The "shape resonances" have been described by Feshbach in elastic scattering cross-section for the processes of neutron capture and nuclear fission [7] in the cloudy crystal ball model of nuclear reactions. These scattering theory is dealing with configuration interaction in multi-channel processes involving states with *different spatial locations.* Therefore these resonances can be called also Feshbach shape resonances. These resonances are a clear well established manifestation of the non locality of quantum mechanics and appear in many fields of physics and chemistry [8,192] such as the molecular association and dissociation processes.

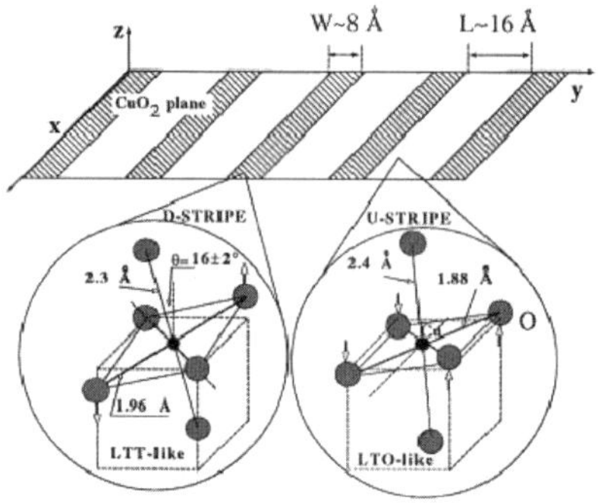

Figure I:2:2. The pictorial view of the superlattice of stripes of mesoscopic lattice fluctuations in La124, and Bi2212 systems determined by EXAFS [87] and resonant anomalous x-ray diffraction [88]

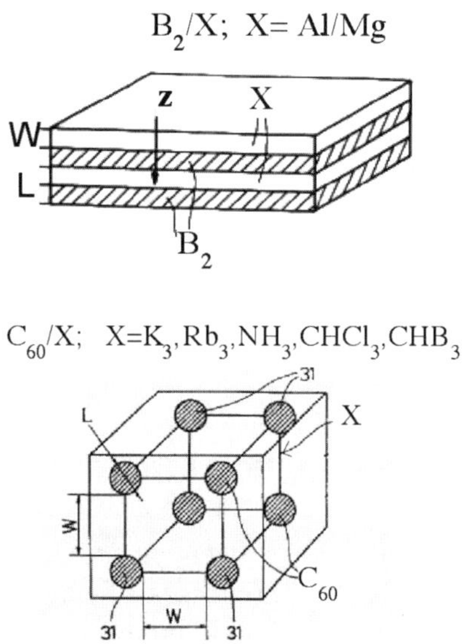

Figure I:2:3. The metal heterostructures at the atomic limit: a superlattice of superconducting layers and a superlattice of superconducting spheres

Feshbach resonances for molecular association and dissociation have been proposed for the manipulation of the interatomic interaction in ultracold atomic gases. In fact the interparticle interaction shows resonances tuning the chemical potential of the atomic gas around the energy of a discrete level of a biatomic molecule controlled by an external magnetic field [193]. This quantum phenomenon has been used by Ketterle to achieve the Bose-Einstein condensation (BEC) in the dilute bosonic gases of alkali atoms [194].Feshbach resonances have recently been used to get a BCS-like condensate in fermionic ultra-cold gases with large values of T_c/T_F [195].

The process for increasing T_c by a Feshbach "shape resonance" was first proposed by Blatt and Thompson [15,16,19] in 1963 for a superconducting thin film. The shape resonance described by Blatt occurs in a superconducting thin film of thickness L where the chemical potential crosses the bottom E_n of the n-th subband of the film, a quantum well, characterized by $k_z = n\pi/L$ with n>1. Therefore it occurs where the chemical potential E_F is tuned near the critical energy $E_F=E_n$ for a 2.5 Lifshitz electronic topological transition (ETT) [191] of type (I) as shown in Fig. 1. At this ETT a small Fermi surface of a second subband disappears while the large 2D Fermi surface of a first subband shows minor variations. In the "clean limit" the single electrons cannot be scattered from the nth to the (n-1)th subband and viceversa, but configuration interaction between pairs in different bands is possible in an energy window around $E_F=E_n$. Therefore the Feshbach shape resonance occurs by tuning the Lifshitz parameter $z=E_F-E_n$ around z=0. In the Blatt proposal z is tuned by changing the film thickness. The prediction of Blatt and Thompson of the oscillatory behavior of T_c as a function of film thickness L has been recently confirmed experimentally for a superconducting film [196] although phase fluctuations due to the electron confinement in the two dimension is expected to reduce the critical temperature.

In 1993, following the experimental evidence of nanoscale striped lattice fluctuations in cuprates shown in Fig. 2, we have proposed to increase T_c via a Feshbach shape resonance in a different class of systems: superlattices of superconducting units as shown in Fig. 3 [75,76,80,81,87-89,93-97,102,109,110].

The idea is that the particular heterogeneous architectures formed by a superlattice of metallic superconducting units intercalated by a different material can bestow decoherence evading qualities to the system. The superconducting units can be dots, spheres, wires, tubes, layers, films and some practical realizations of these architectures are shown in Fig. 3: (i) superlattices of fullerene (quantum spheres), or graphene layers (quantum wells) using carbon atoms intercalated by different atoms can be a practical realization of heterostructures at atomic limit; (ii) superlattices of lead layers intercalated by germanium or silicon layers; (iii) superlattices of bcc metallic layers intercalated by fcc rocksalt layers of a different metal rotated by 45 degrees; (iiii) superlattices of superconducting honeycomb monolayers intercalated by hcp metallic monolayers.

The quantum tricks to realize high T_c superconductors are based on the generic feature of the electronic structure of the superlattices: the presence of different subbands where the charge density associated with each subband is non homogenously distributed in the real space and single electron interband hopping is forbidden by symmetry.

First, the disparity and negligible overlap between electron wavefunctions of different subbands suppress the impurity scattering rate that allows multiband superconductivity in the clean limit. Second, tuning the chemical potential in superlattices of metallic layers shown in Fig. 3 it is possible to approach the type (II) ETT characterized by the opening or closing of a neck in the one of the Fermi surfaces. This simple case is shown in Fig. 1 where the chemical potential is tuned in the region where one Fermi surface changes from a the three-dimensional (3D), for $E_F > E_c$ to a two-dimensional (2D) topology, for $E_F < E_c$.

3. SHAPE RESONANCE IN SUPERLATTICES OF QUANTUM WIRES

The calculation of a Feshbach shape resonance has been carried out for a 2D superlattice of carbon nanotubes of period λ_p on a 2D x,y plane shown in Fig. 4. The electronic structure is similar to the case of a superlattice of stripes [93-96,102] and this type of heterostructures at atomic limit can be classified as “superlattices of quantum wires". While the charge carriers move as free charges in the x direction, the wire direction, they have to overcome a periodic potential barrier V(x,y), with period λ_p, amplitude V_b

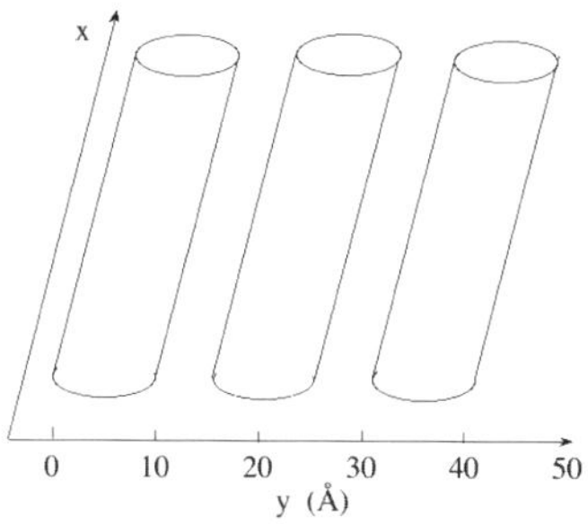

Figure I:2:4. The pictorial view of a 2D superlattice of carbon nanotubes with a period λ_p =1.55 nm in the y direction

and width W along the y direction, constant in the x direction, expressed for x=constant as:

$$V(y) = -V_b \theta\left(\frac{L}{2} - \tilde{y}\right) \quad where \quad \tilde{y} = y - q\,\lambda_p - \frac{\lambda_p}{2} \tag{4}$$

and q is the integer part of y/λ_p.

The solution of the Schrödinger equation for this system, $-\frac{\hbar^2}{2m}\nabla^2\psi(x,y) + V(x,y)\,\psi(x,y) = \mathrm{E}\,\psi(x,y)$ is given as

$$\psi_{n,k_x,k_y}(x,y) = e^{ik_x x} \cdot e^{ik_y q\lambda_p}\,\psi_{n,k_y}(y) \tag{5}$$

where in the stripe

$$\psi_{n,k_y}(y) = \alpha e^{ik_w\tilde{y}} + \beta e^{-ik_w\tilde{y}} \quad for\,|\tilde{y}| < L/2$$

$$k_w = \sqrt{2m_w\left(E_n\left(k_y\right) + V_b\right)/\hbar^2}$$

and in the barrier

$$\psi_{n,k_y}(y) = \gamma e^{ik_b\tilde{y}} + \delta e^{-ik_b\tilde{y}} \quad for|\tilde{y}| \geq L/2$$

$$k_b = \sqrt{2m_b E_n(k_y)/\hbar^2}$$

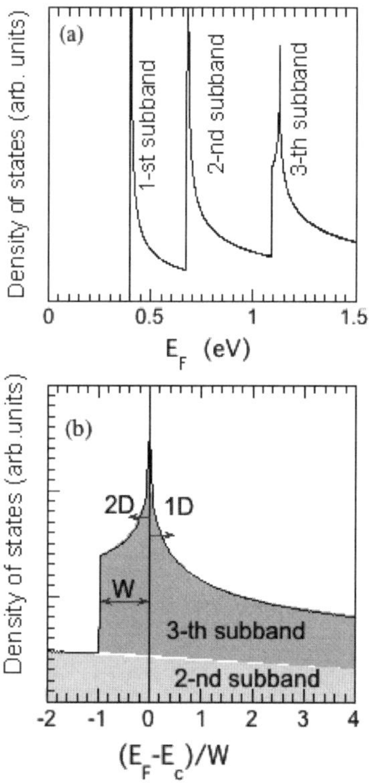

Figure I:2:5. Panel (a) The total density of states (DOS) of a superlattice of nanotubes. The partial DOS of each subband n=1,2,3 gives a peak near the bottom of each subband. Panel (b) shows the details of the DOS near the bottom of the third subband as function of the reduced Liftshitz parameter $"z" = (E_F - E_c)/W$ where W=36.6 meV is the dispersion of the third subband in the y direction of the superlattice, transversal to the nanotube direction. The type (I) ETT occurs at the subband edge ("z"=-1) where the partial DOS of the third subband gives the step-like increase of the DOS. The type (III) ETT occurs at "z"=0 where the DOS shows the main peak

The coefficients α, β, γ and δ are obtained by imposing the Bloch conditions with periodicity λ_p, the continuity conditions of the wave function and its derivative at L/2, and finally by normalization in the surface unit. The solution of the eigenvalue equation for E gives the electronic energy dispersion for the n-th subband with energy

$$\varepsilon_n(k_x,k_y)=\varepsilon(k_x)+E_n(k_y) \text{ where } \varepsilon(k_x)=(\hbar^2/2m)\ k_x^2$$

is the free electron energy dispersion in the x direction and $E_n(k_y)$ is the dispersion in the y direction.

There are N_b solutions for $E_n(k_y)$, with $1 \leq n \leq N_b$, for each k_y in the Brillouin zone of the superlattice giving a dispersion in the y direction of the N_b subbands with $k_x=0$.

The partial density of states (DOS) of the n-th subband gives a step-like increase of the total DOS when the chemical potential reaches the bottom of the subband n=1,2,3 where a type (I) ETT occurs as shown in Fig. 5. The DOS peaks in Fig. 5 are due to partial DOS of each subband tuning the chemical potential by electron doping at the vHs associated with the type (III) ETT in each subband. The panel (b) of Fig. 5 shows the details of the DOS near the type (III) ETT at E_c in the third subband as a function of the reduced Liftshitz parameter $"z"=(E_F-E_c)/W$ where W is the dispersion of the third subband in the y direction of the superlattice, transversal to the nanotube direction.

The pictorial view of the Fermi surfaces of the third and second subbands is shown in Fig. 6. The variation of the Fermi surface topology going through the type (III) ETT is shown. The Fermi surface of the third subband changes from the 1D topology to 2D going from $E_F>E_c$ to $E_F<E_c$ respectively.

The superlattice with its characteristic wavevector $q=2\pi/\lambda_p$ induces a relevant <u>k dependent</u> interband pairing interaction $V_{n,n'}(k,k')$. This is the non BCS interband effective pairing interaction (of any repulsive or attractive nature [14,26]) with a generic cutoff energy $\hbar\omega_0$.

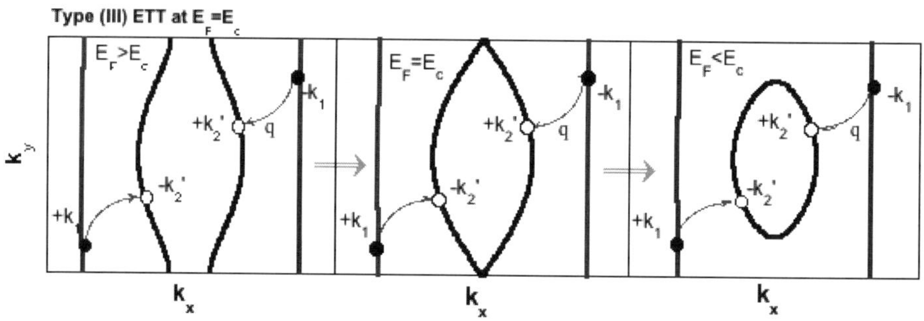

Figure I:2:6. The Fermi surface of the second (red) and third subband (black) of a 2D superlattice of quantum wires near the type (III) ETT where the third subband changes from the one-dimensional (left panel) to two-dimensional (right panel) topology. Going from the left panel to the right panel the chemical potential E_F crosses a vHs singularity at E_c associated with the change of the Fermi topology going from $E_F>E_c$ to $E_F<E_c$, while the Fermi surface of the second subband retains its one-dimensional (1D) character. A relevant inter-band pairing process with the transfer of a pair from the first to the second subband and viceversa is shown

The interband interaction is controlled by the details of the *quantum superposition of states corresponding to different spatial locations i.e,* between the wave functions of the pairing electrons in the different subbands of the superlattice

$$V_{n,n'}(k,k')=V^{o}_{n,k_y;n',k_y'}\theta(\hbar\omega_0-|\varepsilon_n(k)-\mu|)\,\theta(\hbar\omega_0-|\varepsilon_{n'}(k')-\mu|) \qquad (6)$$

where $k=(k_x,k_y)$ and

$$V^{o}_{n,k_y;n',k_y'}=-J\int_S dx\,dy\,\psi_{n,-k}(x,y)\,\psi_{n',-k'}(x,y)\,\psi_{n,k}(x,y)\,\psi_{n',k'}(x,y)$$

$$=-J\int_S dx\,dy\left|\psi_{n,k}(x,y)\right|^2\left|\psi_{n',k'}(x,y)\right|^2$$

n and n' are the subband indexes. k_X (k_X') is the component of the wavevector in the wire direction (or longitudinal direction) and k_y, (k_y') is the superlattice wavevector (in the transverse direction) of the initial (final) state in the pairing process, and μ is the chemical potential.

In the separable kernel approximation, the gap parameter has the same energy cut off $\hbar\omega_o$ as the interaction. Therefore it takes the values Δ_n (k_y) around the Fermi surface in a range $\hbar\omega_o$ depending from the subband index and the superlattice wavevector k_y

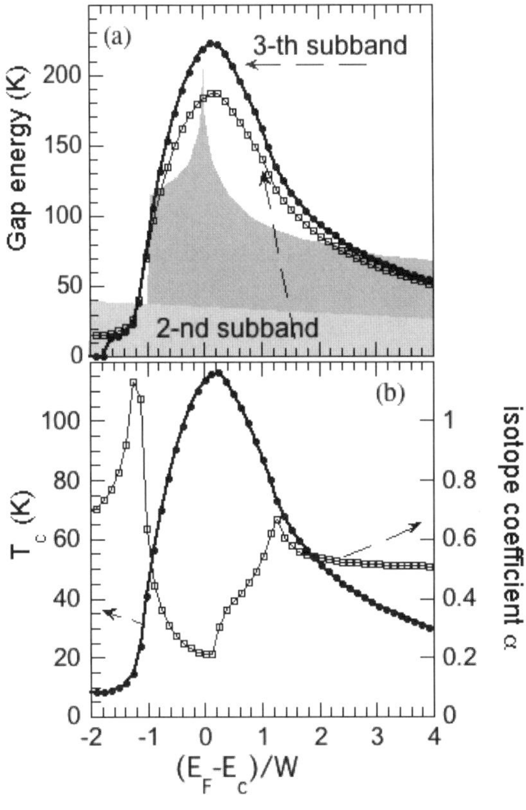

Figure I:2:7. Panel (a): The superconducting gaps in the second, Δ_2, and third, Δ_3, subband in a superlattice of quantum wires as a function of the reduced Lifshitz parameter "z". Panel (b) The critical temperature T_c and the isotope coefficient α at the shape resonance as a function of "z"

The self consistent equation, for the ground state energy gap Δ_n (k_y) is:

$$\Delta_n(\mu, k_y) = -\frac{1}{2N}\sum_{n' k_y' k_x'} \frac{V_{n,n'}\left(k,k'\right)\cdot \Delta_{n'}(k_y')}{\sqrt{(E_{n'}(k_y') + \varepsilon_{k_x'} - \mu)^2 + \Delta_{n'}^2(k_y')}} \tag{7}$$

where N is the total number of wavevectors. Solving iteratively this equation gives the anisotropic gaps dependent on the subband index and weakly dependent on the superlattice wavevector k_y. The structure in the interaction gives different values for the gaps Δ_n giving a system with an anisotropic gaps in the different segments of the Fermi surface.

The superconducting gaps in the second, Δ_2, and third, Δ_3, subband in a superlattice of quantum wires as a function of the reduced Lifshitz parameter "z" are shown in Fig. 7a, where E_F is tuned at E_c for "z"=0, the energy cut off for the pairing interaction is fixed at 500K. The increase of the gap Δ_2 is driven only by the Feshbach resonance in the interband pairing since the partial DOS of the second subband has not peaks.

The critical temperature T_c of the superconducting transition can be calculated by iterative method

$$\Delta_n(k) = -\frac{1}{N}\sum_{n' k'} V_{nn'}(k,k')\frac{\mathrm{tgh}\left(\frac{\xi_{n'}(k')}{2T_c}\right)}{2\xi_{n'}(k')}\Delta_{n'}(k') \tag{8}$$

where $\xi_n(k)=\varepsilon_n(k)-\mu$.

The interband pairing term enhances T_c [93-97,102] by tuning the chemical potential in an energy window around the Van Hove singularities, "z"=0, associated with a change of the topology of the Fermi surface from 1D to 2D (or 2D to 3D) of one of the subbands of the superlattice in the clean limit.

The critical temperature T_c and the isotope coefficient α at the shape resonance are shown in Fig. 7. The result shows the characteristic feature of the T_c amplification by a shape resonance at the maximum critical temperature $T_{c,max}$ the isotope coefficient is close to zero or negative, and shows large positive values up to 1.2 much larger that the standard BCS value α=0.5 at the border of the resonance.

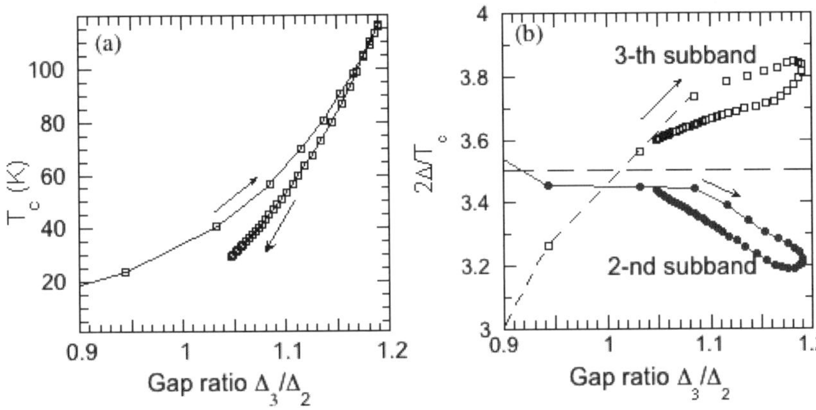

Figure I:2:8. Panel (a) The variation of T_c as a function of the gap ratio $R = \Delta_3/\Delta_2$ at the shape resonance. The maximum T_c is reached for the maximum gap ratio. Panel (b) the ratio $2\Delta_2/T_c$ and $2\Delta_3/T_c$ as a function of the gap ratio Δ_3/Δ_2 . The largest deviation from the single band BCS value $2\Delta/T_c = 0.35$ is reached at the highest gap ratio indicating the key effect of multiband superconductivity

In Fig. 8 we report the variation of T_c as a function of the gap ratio Δ_3/Δ_2 showing that the critical temperature increases by increasing the gap ratio. This provides direct evidence that is a measure of the relevance of the shape resonance in interband pairing. Also the ratio $2\Delta_2/T_c$ and $2\Delta_3/T_c$ show large deviations from the BCS value $2\Delta/T_c = 0.35$ for a single band indicating the key effect of multiband superconductivity. These calculations show that the interband pairing enhances T_c [93, 95] by tuning the chemical potential at the shape resonance.

4. SHAPE RESONANCES IN SUPERLATTICES OF QUANTUM WELLS

MgB_2 provides the simplest high T_c superconductor therefore it could play a key role for understanding high T_c superconductivity as atomic hydrogen for quantum mechanics. There is now growing evidence that MgB_2 is a practical realization of the proposed T_c amplification process driven by Feshbach shape resonances in interband pairing [116,120,132,139,144,162, 176,177,180] representing the case of a superlattice of quantum wells

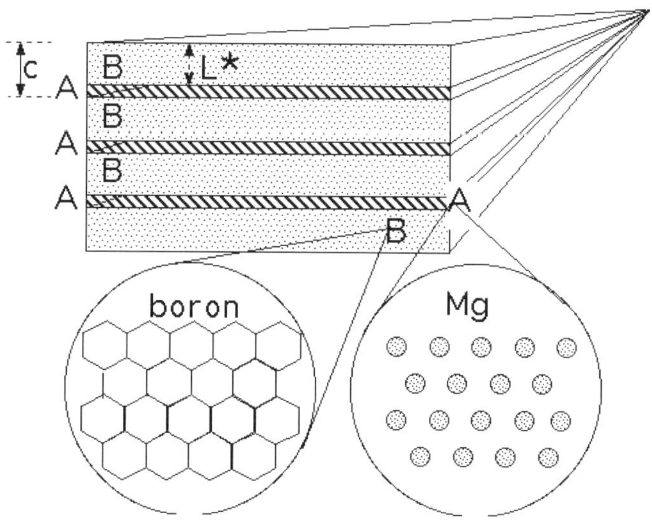

Figure I:2:9. The superlattice of metallic layers (B) made of graphite-like honeycomb boron planes separated by hexagonal Mg layers (A) forming a superlattice ...ABABAB... where the c-axis is the period of the superlattice

The light-metal diborides AB_2 (A=Mg, Al) are not common natural compounds. They have been discovered as residuals in the chemical processing [197,198] for the reduction of boron oxide with electropositive metals to obtain elemental boron. Following the synthesis and characterization of aluminum diboride (AlB_2) in 1935 [199] magnesium diboride (MgB_2) was synthesized in 1953 [200-201] when the chemical interest in borides was driven by nuclear power industry for control rods and neutron shields.

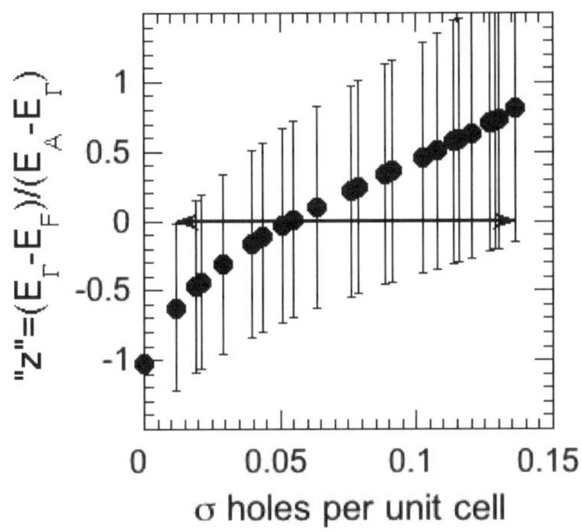

Figure I:2:10. The reduced Lifshitz parameter $"z" = (E_\Gamma - E_F)/(E_A - E_\Gamma)$, where $(E_A - E_\Gamma)$ is the full energy band dispersion in the c-axis direction, as a function of the number of holes in the σ subband in Al doped MgB_2. The quantum uncertainty in the z value is indicated by the error bars that are given by $D\sqrt{\langle\sigma^2_{x,y}\rangle}/(E_A - E_\Gamma)$ where D is the deformation potential and $\langle\sigma^2_{x,y}\rangle$ is the mean square boron displacement at T=0K associated with the E_{2g} mode measured by neutron diffraction [139]. The T_c amplification by Feshbach shape resonance occurs in the σ hole density range shown by the double arrow indicating where the 2D-3D ETT sweeps through the Fermi level because of zero point lattice motion, i.e., where the error bars intersect the z=0 line

For many years MgB_2 was not considered as a possible superconductor by the scientific community on the basis of conventional theories or material science rules for search of high T_c superconductors. The AlB_2 crystalline structure is a heterostructure at the atomic limit made of superconducting layers (boron monolayers) intercalated by different layers (Al or Mg hcp monolayers) shown in Fig. 9.

The σ band is usually fully occupied in non superconducting diborides where there are no holes in the σ band, so these diborides, as AlB_2 have a Fermi surface of the type shown for the case $E_F>E_A$ in Fig. 10. In MgB_2 the σ band is partially unoccupied due to the electron transfer from the boron layers to the magnesium layers. In fact the chemical potential E_F in MgB_2 is at about 750 meV below the energy E_A of the top of the σ band. Moreover the chemical potential E_F in MgB_2 is also at about 350 meV below the energy of the Γ point in the band structure ($E_\Gamma>E_F$). Therefore the σ Fermi surface of MgB_2 where $E_F<E_\Gamma<E_A$ has the corrugated tubular shape with a two-dimensional topology of the type shown in Fig. 10 for the case $E_F<E_\Gamma$. Going from below ($E_F<E_\Gamma<E_A$) to above the energy of the Γ point the σ Fermi surface becomes a closed Fermi surface with 3D topology like in $AlMgB_4$ that belongs to the Fermi surface type for the case $E_\Gamma<E_F<E_A$.Therefore by tuning the chemical potential E_F, by electron doping the σ subband, it is possible to reach the point where E_F is tuned at the 2D/3D Van Hove singularity. This is a type (II) 2.5 Lifshitz electronic topological transition (ETT) with the disruption of a "neck" in the σ Fermi surface with the critical point at $E_F=E_\Gamma$. The changes of physical properties near the ETT transition are studied here as a function of the reduced Lifshitz parameter "z"=(E_Γ-E_F)/(E_A-E_Γ), where W=(E_A-E_Γ) is the energy dispersion in the c-axis direction due to electron hopping between the boron layers (W=0.4 eV in MgB_2 but it changes with chemical substitutions). The influence of the proximity to a type (II) electronic topological transition on the anomalous electronic and lattice properties of MgB_2 is shown by the anomalous pressure dependence of the E_{2g} phonon mode and T_c (202, 203).
The response of the superconducting properties of diborides to Fermi level tuning has been studied by electron doping using atomic substitutions, in fact it is possible to reduce the number of holes in the σ band from about 0.15 holes per unit cell in MgB_2 up to reach zero σ holes at the top of the σ band for "z"=-1 where $E_F=E_A$.

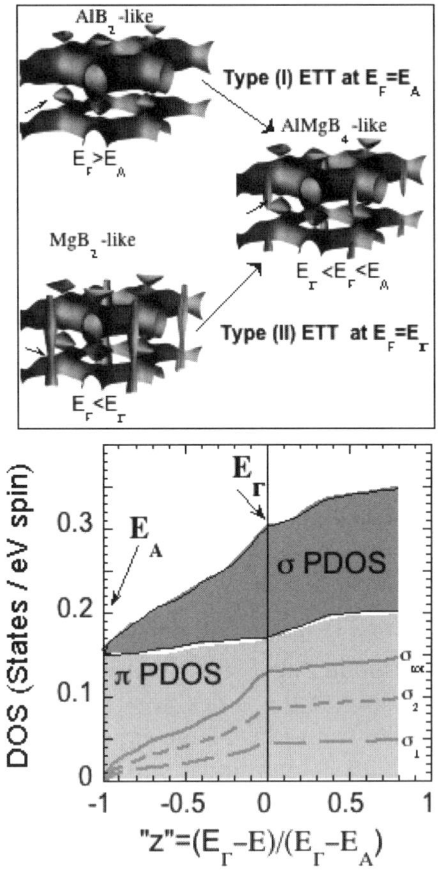

Figure I:2:11. The total density of states (DOS) in MgB_2 and the partial DOS (PDOS) of the σ and π band as a function of the reduced Lifshitz parameter 'z". The upper curve shows two different types of 2.5 Lifshitz electronic topological transition (ETT) in electron doped MgB_2 associated with changes of the Fermi surface topology. The type (I) ETT: with the appearance of the closed 3D σ Fermi surface by tuning the Fermi energy at the critical point $E_F=E_A$ where E_A is the energy of the A point in the band structure. The type (II) ETT with the disruption of a "neck" in the σ Fermi surface by tuning the chemical potential E_F at the critical point in the band structure $E_F=E_\Gamma$ where the σ Fermi surface changes from a 2D corrugated tube for $E_\Gamma>E_F$ to a closed 3D Fermi surface for $E_\Gamma<E_F$ while the large π Fermi surface keeps its 3D topology [163]

We report in Fig. 11 the variation of the reduced Lifshitz parameter “z” as a function of the number of holes in the σsubband for the case of Al doped $Mg_{1-x}Al_xB_2$. In order to understand the physics of super-conductivity in diborides we have to consider that the Lifshitz parameter “z” has an intrinsic uncertainty due to quantum fluctuations. The error bars in Fig. 11 indicate the quantum uncertainty in the “z” value induced by zero point lattice fluctuations. In fact the in-planeboron phonon mode with E_{2g} symmetry near the zone-center, split the partially occupied σ bands. The energy splitting of the two degenerate σbands is given by $\Delta E=D\sqrt{\langle\sigma^2_{x,y}\rangle}$ where D is the deformation potential and $\sqrt{\langle\sigma^2_{x,y}\rangle}$ is the mean square in plane displacement at T=0K of the boron atoms due to zero point motion measured by neutron scattering [132,139]. The energy splitting is $\Delta E=\pm0.4$ eV in MgB_2 and it decreases a little going to $AlMgB_4$ where there are about 0.02 σholes per unit cell. Therefore the quantum lattice fluctuations induce the electronic quantum fluctuations of the Γpoint relative to the Fermi level, i.e., of the Lifshitz parameter z.

This induces quantum charge fluctuations between the two σ bands and between the σ and π bands involving electronic states within the energy window of ±0.4 eV around the Fermi level. These charge fluctuations control the energy window of the states involved in the pairing processes therefore we can estimate from Fig. 11 the expected width of the Feshbach shape resonance. The high T_c amplification by Feshbach shape resonance should occur in the range $-0.8<$ "z" $<+0.8$ where the 2D/3D Van Hove singularity sweeps through the Fermi level that is represented in Fig. 11 where the value z=0 follows within the error bars of z.

The Lifshitz parameter z as a function of x in the case of the Al and Sc substitutions for Mg in the $Mg_{1-x}Al_xB_2$, and $Mg_{1-x}Sc_xB_2$ systems and for the C for B substitution in the $MgB_{2-x}C_x$ system has been calculated by R. De Coss et al. by band structure calculations described elsewhere [204] therefore it has been possible to convert the variation of the critical temperature as a function of the number density of substituted ions x to the variation of T_c versus the universal reduced Lifshitz parameter “z” for all doped magnesium diborides.

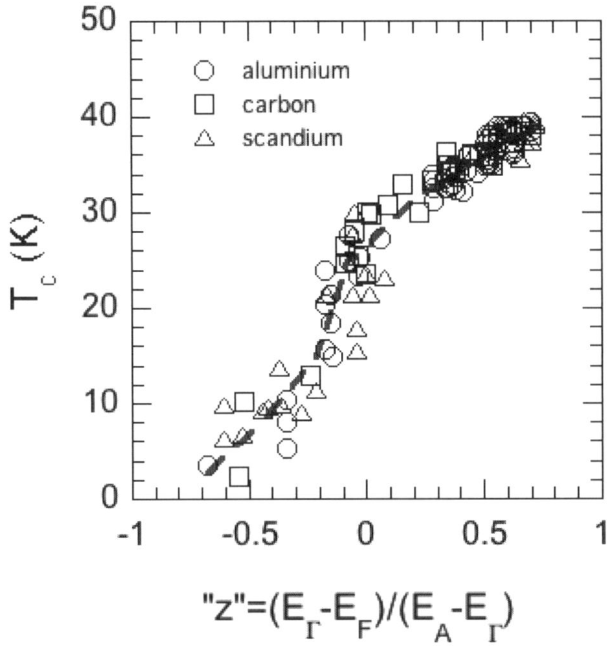

Figure I:2:12. The critical temperature in aluminum, scandium and carbon doped magnesium diborides as a function of the Lifshitz parameter "z"= (E_Γ-E_F)/(E_A-E_Γ)

The critical temperature T_c obtained by aluminium [143], carbon [183,184] and scandium [177] substitutions are reported in Fig. 12. The universal scaling of the critical temperature T_c, versus "z" show the negligible effects of impurity scattering in different substitutions because of the suppression of interband impurity scattering in the superlattice.

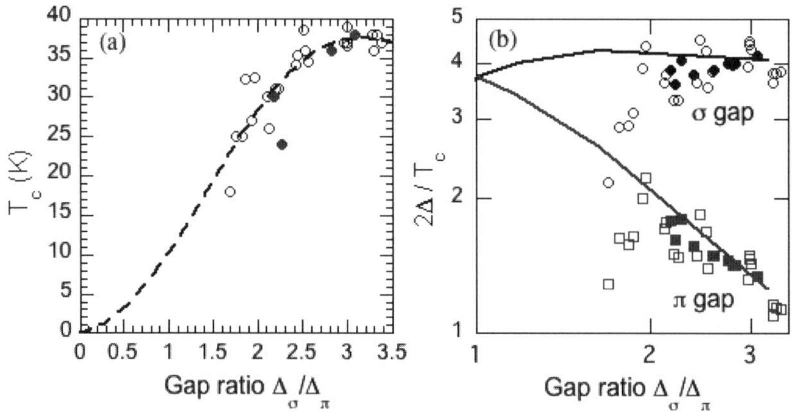

Figure I:2:13. Panel a): The experimental T_c in C for B (filled symbols) [183] and in Al for Mg (empty symbols) substituted MgB_2 as a function of the gap ratio Δ_σ/Δ_π . Panel b): The ratio $2\Delta/T_c$ for the σ gap (circles) and for the π gap (squares) as a function of the gap ratio Δ_σ/Δ_π . The solid lines are the theoretical prediction (published before the experiments have been done [162]) where the effect of atomic substitutions in this superlattice of boron layers do not introduce interband impurity scattering effects but only tune the chemical potential toward the 2D/3D vHs in the Feshbach shape resonance range

We report in Fig. 13 the superconducting critical temperature T_c obtained by aluminium [143], carbon [183,184] and scandium [177] substitutions as a function of the superconducting gap ratio Δ_σ/Δ_π. The experimental results clearly show that the critical temperature increases by increasing the gap ratio as it was shown by theoretical calculations for a superlattice of wires in Fig. 8. In Fig. 13 we report the experimental ratio $2\Delta_\sigma/T_c$ and $2\Delta_\pi/T_c$ for the π and in the σ band as a function of the gap ratio. The highest T_c occurs for the largest deviation from the BCS value 3.5

These results show that the two gaps behaviour is present over all the range of the reduced Lifshitz parameter -0.8<"z"<+0.8. These results support the predictions that the doped materials remain in the clean limit for interband pairing although the large number density of impurity centres. This results falsifies the predictions of the multigap suppression because of impurity scattering due to substitutions.

In order to identify the Feshbach shape resonance we have plotted in Fig. 14 the ratio T_c/T_F(exp) for the aluminum doped case where T_F is the Fermi temperature $T_F=\varepsilon_F/K_B$ and $\varepsilon_F=E_A-E_F$ is the Fermi energy of the holes in the σ band, and T_c is the measured critical temperature. The T_c/T_F ratio is a measure of the pairing strength $(k_F\xi_0)^{-1}$ where k_F is the Fermi wavevector and ξ_0 is the superconducting coherence length. In fact in the single band BCS theory this ratio is given by $T_c/T_F = 0.36/k_F\xi_0$.

.

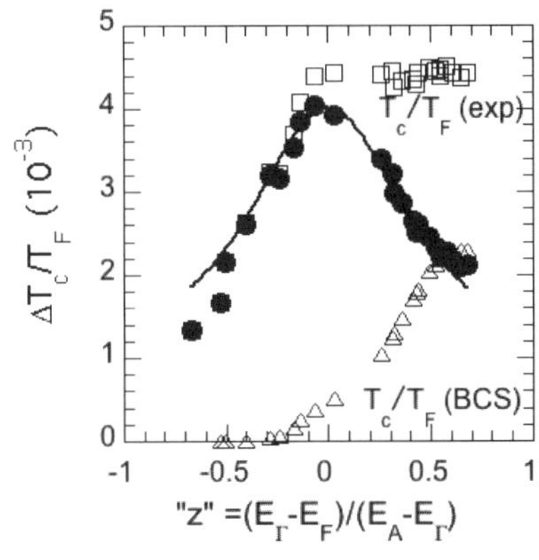

Figure I:2:14. The ratio T_c/T_F(exp) (open squares) for the aluminium doped case where T_F is the Fermi temperature $T_F=E_F/K_B$ for the holes in the σ band and T_c is the measured critical temperature for Al doped samples as a function of the reduced Lifshitz parameter "z". The expected ratio T_c/T_F(BCS) (open triangles) for the critical temperature in a BCS single s band, The ratio $\Delta T_c/T_F = (T_c(\mathrm{exp}) - T_c(BCS))/T_F$ (solid dots) that measures the increase due to the interband pairing, for the Feshbach shape resonance, fitted with a asymmetric Lorentzian with Fano line-shape centred at z=0, a half width Γ/2=0.47 and asymmetry parameter q=5

In Fig. 14 we have plotted also the expected ratio T_c/T_F(BCS) for the critical temperature in the single σ band of MgB_2 calculated by the standard BCS approximations using the McMillan formula, the density of states and electron phonon coupling obtained by band structure theory that gives for MgB_2 $T_c(BCS)$=20K [148].

The increase of the critical temperature determined by the Feshbach shape resonance [118,120,205] of the interband pairing is given by $\Delta T_c/T_F = (T_c(\exp) - T_c(BCS))/T_F$ that is plotted in Fig. 14 as a function of Lifshitz parameter "z". It has been fitted with an asymmetric Lorentzian with Fano line-shape [8] centred at z=0, with the half width Γ/2=300meV and an asymmetry parameter q=4. This experimental curve provides the clear experimental evidence for the T_c enhancement driven by interband pairing shows a resonance centred at z=0, as expected for Feshbach shape resonanceIn conclusion we have discussed the Feshbach shape resonances around a type (I), type (II) and type (III) ETT in heterostructures and in the particular case of MgB_2 where the two band scenario is well established [205-207], we have shown that the Feshbach shape resonance in interband pairing occurs in a superlattice of superconducting layers near a type (II) ETT in this particular multiband superconductor in the clean limit. The Feshbach shape resonance increases the critical temperature from low T_c diborides up to T_c =40K in MgB_2 and it provides the most clear case of T_c amplification beyond the limit of 20K. These results confirm that the Feshbach shape resonance in interband pairing could be a possible mechanism for high T_c i.e., the quantum trick for suppressing decoherence effects at high temperature.

We finally summarize the physical conditions to get a Feshbach shape resonance suppressing decoherence effects of temperature and driven by quantum superposition of pairs in states corresponding to different spatial locations and different parity. *First*, the material architecture is made of a superlattice of superconducting units (B) at atomic limit intercalated by a different material (Al, Mg) like in AlB_2 diborides; *second*, the chemical potential is tuned in an energy window around a 2.5 Lifshitz topological electronic transition associated with the change of the topology of theFermi surface of one of the subbands; *third*, the energy window of the resonance is controlled by quantum fluctuations driven by the zero point lattice vibrations that sweep the Van Hove singularity through the Fermi level.

ACKNOWLEDGMENTS

This work is supported by MIUR in the frame of the project Cofin 2003 "Leghe e composti intermetallici: stabilità termodinamica, proprietà fisiche e reattività" on the "synthesis and properties of new borides".

REFERENCES

1. J.G. Bednorz and K.A. Müller, *Rev. Mod. Phys.* 60, 565 (1988).
2. G. Careri *Order and Disorder in Matter* Addison-Wesley, NY (1984); P. Davies *New Scientist,* 2477 (11 December 2004).
3. J.R. Schrieffer, *Phys. Today* 45, 46 (1992); A. Leggett in *The New Physics* P. Davies ed., Cambridge University Press pp. 268-288 (1989); B. D. Josephson *Phys. Lett.* 1, 251 (1962); J. Bardeen, L. N. Cooper and J. R. Schrieffer, *Phys. Rev.* 108, 1175 (1957); F. London and H. London, *Proc. Roy. Soc. (London)* Ser **A152**, 24 (1935).
4. F. London *Superfluids*, Vol. 2 J. Wiley and Sons, New York, (1954); R.P. Feynman *Progress In Low Temperature Physics,* C.J. Gorter ed. **1**, 17, North Holland, Amsterdam (1955);
5. N.N. Bogulyubov *J. Physics (USSR)* 11, 23 (1947); L. D. Landau *J. Phys. USSR* 5, 71 (1941); ibidem 11, 91 (1947).
6. S. Tomonaga, *Prog. Theo. Phys. (Kyoto).* 2, 6 (1947).
7. H. Feshbach, C.E. Porter, and V.F. Weisskopf *Phys. Rev.* 96, 448 (1954); H. Feshbach, *Ann. Phys. N.Y.* 5, 357 (1958).
8. U. Fano *Nuovo Cimento* 128, 156 (1935); condmat/ 0502210; U. Fano, Phys. Rev. 124, 1866 (1961); Ugo Fano *Atomic Collisions and Spectra* Elsevier, Amsterdam (1986).
9. N.N. Bogolyubov *Soviet Phys. JETP* 7, 41 (1958); *Nuovo Cimento* 7, 794 (1958); N.N. Bogolyubov, V.V. Tolmachev, and D. V. Shirkov *A New Method in the Theory of Superconductivity"* Izv. Akad. Nauk. SSSR, Moscow (1958). Consultants Bureau, N.Y., (1959).
10. V.A. Moskalenko, *Phys. Met. and Metallog.* 8, 25 (1959).
11. H. Suhl, B.T. Matthias, and L.R. Walker *Phys. Rev. Lett.* 3, 552 (1959).
12. B.T. Matthias *J. Phys. Soc. Japan* 17 Supplement B-I, 104 (1962).
13. M. Suffczynski *Phys. Rev.* 128, 1538 (1962).
14. J. Kondo *Prog. Theor. Phys.* 29, 1 (1963).
15. J.M. Blatt and C.J. Thompson *Phys. Rev. Lett.* 10, 332 (1963).
16. C.J. Thompson and J.M. Blatt, *Phys. Lett.* 5, 6 (1963).
17. J. W. Garland, Jr., *Phys. Rev. Lett.* 11, 111 (1963).
18. N.V. Zavaritskii *Soviet Physics JEPT* 18, 1260 (1963).
19. J.M. Blatt *Theory of Superconductivity* Academic Press New York 1964, (for shape resonances see 362 pp. and 215 p.p.).
20. J.R. Schrieffer, *Theory of Superconductivity*, 1964, (for interband pairing see 300 pp.).

21. T. H. Geballe, *Rev. Mod. Phys.* 36, 134 (1964); D.R. Tilley, *Proc. Phys. Soc.* 84, 573 (1964).
22. B.T. Geilikman, *Sov. Phys.* JEPT 21, 796 (1965).
23. Yu Luh and Gu Ben-Yuan, *Acta Physica Sinica* 21 838 (1965).
24. V.A. Moskalenko, *Sov. Phys. JETP* 24, 780 (1966); V.A. Moskalenko and M. E. Palistrant, *Sov. Phys. JETP* 22, 536 (1966)
25. R. Vasudevan and C.C. Sung, *Phys. Rev.* 144, 237 (1966).
26. A. J. Leggett, *Prog. Theor. Phys.* 36, 901 (1966).
27. A. J. Leggett, *Prog. Theor. Phys.* 36, 931 (1966).
28. B.T. Geilikman, R. O. Zaitsev, and V. Z. Kresin, *Sov. Phys. Solid State* 9, 642 (1967).
29. L.Z. Kon, *Phys. Met. Metallogr.* (USSR). 23, (1967).
30. J.W. Garland, Jr., *Phys. Rev.* 153, 460 (1967).
31. V.C. Wong, and C.C. Sung, *Phys. Rev. Lett.* 19, 1236 (1967); C.C. Sung and V.C. Wong *J. Phys. Chem. Solids* 28, 1933 (1967); W. S. Chow, *Phys. Rev.* 172, 467 (1968); T. Kusakabe, Prog. Theor. Phys. 43, 907 (1970)
32. M.E. Palistrant and M.K. Kolpajiu, *Phys. Lett. A* 41, 123 (1972).
33. A.G. Aronov and E.B. Sonin, *JEPT Lett.* 36, 556 (1973).
34. V. Kresin *J Low Temp Phys* 11, 519 (1973).
35. B.T. Geilikman and V. Kresin in *Kinetic and non stationary phenomena in superconductors* J. Wiley, New York, pp. 34 and 80 (1974)
36. P. Entel, and M. Peter, *J. Low Temp. Phys.* 22, 613 (1975).
37. P. Entel, W. Klose, O. Fischer, and G. Bongi, *Z. Phys. B* 21, 363 (1975).
38. P. Entel, and M. Peter, *Jour. of Low Temp. Phys.* 22, 613 (1976).
39. H. Butler, and P.B. Allen, in *Superconductivity in d and f-band Metals*, edited by D. H. Douglass, Plenum, New York, 1976.
40. P.B. Allen, Phys. Rev. B 17, 3725 (1978); P.B. Allen and B. Mitrovic, in: Solid State Physics, F. Seitz, D. Turnbull, and H. Ehrenreich editors, Academic, New York, p. 1 (1982).
41. G. Vignale and K.S. Singwi, *Phys. Rev. B* 31, 2729 (1985).
42. J. Ranninger, and S. Robaszkiewicz, *Physica (Amsterdam)* 53B, 468 (1985).
43. D.H. Lee, and J. Ihm, *Solid State Commun.* 62, 811 (1987).
44. K. Yamaij, and A. Abe, *J. Phys. Soc. Japan* 56, 4237 (1987).
45. K. Yamaij, *Sol. State Commun.* 64, 1157 (1987).
46. S. Robaszkiewicz, R. Micnas, and J. Ranninger, *Phys Rev B* 36, 180 (1987).
47. V. Kresin and S. Wolf in "Novel Superconductivity" S. Wolf and V. Kresin editors Plenum, New York (1987); N.W. Ashcroft, ibidem p. 301 (1987).
48. V. Kresin *Sol. State Commun.* 63, 725 (1987); V. Kresin, and S. Wolf *Sol. State Commun.* 63, 1141 (1987)
49. J. Ruvalds *Phys. Rev. B* 35, 8869 (1987
50. Ji-hai Xu, *Solid State Commun.* 65, 135 (1988).
51. P. Konsin, N. Kristoffel, and T. Ord, *Phys. Lett.* A 129, 399 (1988).
52. O. Entin-Wohlman, and Y. Imry, *Physica C* 153-155, 1323 (1988).
53. O. Entin-Wohlman, and Y. Imry, *Phys. Rev. B* 40, 6731 (1989).

54. A.N. Vasilchenko, and A.V. Sokol, *Zh. Eksp. Teor. Phys.* 96, 377 (1989).
55. V.A. Moskalenko, M.E. Palistrant, V.M. Vakalyuk, and I.V. Padure, *Solid. State. Commun.*, 69, 747 (1989).
56. E.G. Volovik, *JETP Lett.* 49, 790 (1989).
57. R. Friedberg, and T.D. Lee, *Phys. Rev.* B40, 6745 (1989).
58. U. Hofmann, J. Keller, and M. L. Kulic, *Z. Phys.* 81, 25 (1990).
59. K. Yamaji, *J. Phys. Soc. Jpn.* 59, 677 (1990).
60. V. Kresin, and S. Wolf *Physica C* 169, 476 (1990); V. Kresin, and S. Wolf, *Phys Rev* B 41, 4278 (1990).
61. A.A. Abrikosov, Physica C 182, 191 (1994); A.A. Abrikosov and R.A. Klemm, *Physica C* 191, 224 (1992)
62. V.A. Moskalenko, M.E. Palistrant, and V.M. Vakalyuk, *Sov. Phys. Usp.* 34, 717 (1991).
63. J.E. Hirsh, and F. Marsiglio, *Phys. Rev. B* 43, 424 (1991).
64. F. Marsiglio, *Jour. of Low Temp. Phys.* 87, 659 (1992).
65. V.Z. Kresin, and S. A. Wolf, *Phys. Rev. B* 46, 6458 (1992).
66. V. Kresin, S. Wolf, and G. Deutscher *J Phys C* 191, 9 (1992).
67. E. Langmann, *Phys. Rev. B* 46, 9104 (1992).
68. N. Bulut, D. J. Scalapino, and R.T. Scalettar, *Phys. Rev. B* 45, 5577 (1992).
69. A. Bianconi *"High T_c superconductors made by metal heterostuctures at the atomic limit"* European Patent N. 0733271 (priority date: 07 12 1993).
70. A. Bianconi "*Process of increasing the critical temperature T_c of a bulk superconductor by making metal heterostructures at the atomic limit*" United State Patent: No.:US 6, 265, 019 B1. (priority date: 07 12 1993).
71. A. Bussmann-Holder, L. Genzel, A. Simon, and A.R. Bishop *Z. Phys. B* 91, 271 (1993); ibidem 92, 149 (1993).
72. B.K. Chakraverty, *Phys Rev B* 48, 4047 (1993).
73. I.I. Mazin, A.I. Liechtenstein, C.O. Rodriguez, O. Jepsen, and O.K. Andersen, *Physica C* 209, 125 (1993)
74. L. Genzel, M. Bauer, H.-U. Habermeier, and E.H. Brandt, *Z. Phys. B* 90, 3 (1993).
75. A. Bianconi, *Sol. State Commun.* 89. 933 (1994).
76. A. Bianconi, M. Missori *J. Phys. I (France).* 4, 361 (1994).
77. N. Kristoffel, P. Konsin, and T. Ord, *Rivista Nuovo Cimento* 17, 1 (1994).
78. A.A. Golubov, O.V. Dolgov, E.G. Maksimov, I.I. Mazin, and S.V. Shulga, *Physica C* 235-240, 2383 (1994).
79. K. Yamaji, *Physica* 222C 64, 349 (1994).
80. A. Bianconi, M. Missori, N.L. Saini, H. Oyanagi, H. Yamaguchi. D.H. Ha. and Y. Nishiara *Journal of Superconductivity* 8, 545 (1995).
81. A. Bianconi, M. Missori, N.L. Saini, H.Oyanagi, H. Yamaguchi, H. Oyanagi, H. Yamaguchi, Y. Nishiara, D.H. Ha. and S. Della Longa in *Anharmonic Properties of High T_C Cuprates* edited by D. Mihailovic G. Ruani E. Kaldis. K.A. Müller World Scientific Singapore (1995). pag.127-138
82. G.C. Asomba, *Physica C* 244, 271 (1995); G.C. Asomba, *Physica C* 245, 355 (1995).

83. A.A. Golubov, and I.I. Mazin, *Physica C* 243, 153 (1995).
84. R. Combescot, and X. Leyronas, *Phys. Rev. Lett.* 75, 3732 (1995).
85. W.C. Wu, and A. Griffin, *Phys. Rev. Lett.* 74, 158 (1995).
86. V.Z. Kresin, and S.A. Wolf, *Phys. Rev. B* 51, 1229 (1995).
87. A. Bianconi, N.L. Saini, A. Lanzara, M. Missori., T. Rossetti, H. Oyanagi, H. Yamaguchi, K. Oka, T. Ito, *Phys. Rev. Lett* 76. 3412 (1996).
88. A. Bianconi, N.L. Saini, T. Rossetti, A. Lanzara, A. Perali, and M. Missori, *Phys. Rev. B* 54, 12018 (1996).
89. A. Perali, A. Bianconi A. Lanzara, and N.L. Saini *Sol. State Commun.* 100, 181 (1996); A. Bianconi, N.L. Saini A. Lanzara, A. Perali and A. Valletta in "High T_c Superconductivity 1996, Ten Years after the Discovery" E. Kaldis, E. Liarokapis and K. A. Müller editors, Kluwer, Dordrecht, p. 383 (1996).
90. C.F. Richardson and N.W. Ashcroft *Phys. Rev B* 54, R764 (1996).
91. R. Combescot and X. Leyronas *Phys. Rev B* 54, 4320 (1996)
92. P. Konsin, N. Kristoffell, and P. Rubin, *Solid State Commun.* 97, 567 (1996).
93. A. Bianconi, A. Valletta, A. Perali, and N.L. Saini, *Solid State Commun.* 102, 369 (1997).
94. A. Perali, A. Valletta, G. Bardelloni, A. Bianconi, A. Lanzara, and N.L. Saini, *J. Superconductivity* 10. 355 (1997).
95. A. Valletta, G. Bardelloni, M. Brunelli, A. Lanzara, A. Bianconi, and N.L. Saini, *J. Superconductivity* 10. 383 (1997).
96. A. Valletta, A. Bianconi, A. Perali, N.L. Saini, *Z. Phys. B* 104, 707 (1997).
97. N.L. Saini, J. Avila, A. Bianconi, A. Lanzara, M.C. Asenzio, S. Tajima, G.D. Gu, and N. Koshizuka, *Phys. Rev. Lett.* 79, 3467 (1997).
98. S.D. Adrian, S. A. Wolf, O.V. Dolgov, S. Shulga, V.Z. Kresin, *Phys. Rev. B* 56, 7878 (1997).
99. X. Leyronas, and R. Combescot, *Physica C* 290, 215 (1997).
100. N. Kristoffell, P. Konsin, and P. Rubin, *Phys. Status Solid B* 203, 501 (1997).
101. A.A. Golubov, and I.I. Mazin, *Phys. Rev.* B 55, 15146 (1997).
102. A. Bianconi, A. Valletta, A. Perali, and N.L. Saini, *Physica C* 296, 269 (1998).
103. N. Kristoffell, P. Konsin, and P. Rubin, *Phys. Status Solid B* 208, 145 (1998).
104. R. Combescot, *Phys. Rev. B* 57, 8632 (1998).
105. D. F. Agterberg, V. Barzykin and L. P. Gor'kov *Europhys. Lett.*, 48, 449 (1999)
106. M.L. Kulic, and O.V. Dolgov, *Phys. Rev. B* 60, 13062 (1999).
107. R. Combescot, *Phys. Rev.Lett.* 83, 3766 (1999).
108. M. Imada, and M. Kohno, *Phys. Rev. Lett.* 84, 143 (2000).
109. A. Bianconi, *Int. J. Mod. Phys. B* 14, 3289 (2000); A. Bianconi, S. Agrestini, G. Bianconi, D. Di Castro, and N.L. Saini, in *"Stripes and Related Phenomena"* A. Bianconi, N.L. Saini Editors, Kluwer Academic/Plenum publisher, New York, 2000 p. 9-25.
110. N.L. Saini, and A. Bianconi, *Int. J. Mod. Phys. B,* 14, 3649 (2000).
111. N. Kristoffel, and P. Rubin, *Physica C* 356, 171 (2001).
112. M. Granath, V. Oganesyan, S. A. Kivelson, E. Fradkin, and V. J. Emery, *Phys. Rev. Lett.* 87, 167011 (2001).

113. S. Basu, A. Callan-Jones, and R.J. Gooding, *Phys. Rev. B* 66, 144507 (2002).
114. J. Tanaka, *Phys. Rev. Lett.* 88, 017002 (2002).
115. J. Nagamatsu, N. Nakagawa, T. Muranaka, Y. Zenitani, and J. Akimitsu, *Nature* 410, 63 (2001).
116. A. Bianconi et al., cond-mat/0102410 (22 Feb 2001); S. Agrestini, D. Di Castro, M. Sansone, N.L. Saini, A. Saccone, S. De Negri, M. Giovannini, M. Colapietro, and A. Bianconi, *J. of Phys.: Cond. Matter* 13, 11689 (2001).
117. M. Imada, cond-mat/0103006 (1 March 2001) ; M. Imada, *J. Phys. Soc. Jpn.* 70 1218 (2001).
118. A.V. Shulga, D.-L. Drechsler, H. Eschrig, H. Rosner, and W.E. Pickett cond-mat/0103154 (7 March 2001) ; G. Fuchs, S.-L. Drechsler, S.V. Shulga, A. Handstein, V. Narozhnyi, K. Nenkov, and K.-H. Mueller, in *"High Temperature Superconductors"* Vol. 41, A. Narlikar editor (Nova Science, NY 2002). pp. 171-198.
119. E. Bascones and F. Guinea. Cond-mat/0103190 (8 March 2001).
120. A. Bianconi et al., cond-mat/0103211 (9 March 2001) ; A. Bianconi, D. Di Castro, S. Agrestini, G. Campi, N.L. Saini, A. Saccone. S. De Negri, M. Giovannini, J. Phys.: Condens. Matter 13, 7383 (2001).
121. A. Bianconi, D. Di Castro, S. Agrestini, G. Campi, N.L. Saini, A. Saccone, S. De Negri, M. Giovannini, Presented at APS March meeting, March 12, 2001 Seattle,Washington, "Session on MgB_2" talk 79, http://www.aps.org/MAR01/mgb2/talks.html#talks79
122. H. Rosner, W.E. Pickett, S. Shulga, S.-L. Drechsler, and H. Eschrig, Presented at APS March meeting, March 12, 2001 Seattle,Washington, Session on MgB_2 Superconductivity, talk 68, http://www.aps.org/MAR01/mgb2/talks.html#talks68
123. K. Yamaji, cond-mat/0103431 (21 March 2001); K. Yamaji, *J. Phys. Soc. Jpn.* 70, 1476 (2001).
124. N. Kristoffel, T. Ord, cond-mat/0103536 (26 March 2001); T. Örd, and N. Kristoffel *Physica C* 370, 17 (2002).
125. A.Y. Liu, I.I. Mazin, and J. Kortus, cond-mat/0103570 (27 Mar 2001); A.Y. Liu, I.I. Mazin, and J. Kortus, *Phys. Rev. Lett.* 87, 087005 (2001).
126. X.K. Chen, et al., cond-mat/0104005 (31 March 2001); X.K. Chen, M.J. Konstantinovic, J.C. Irwin, D.D. Lawrie, and J.P. Franck, *Phys. Rev. Lett.* 87, 157002 (2001).
127. F. Bouquet, R. A. Fisher, N. E. Phillips, D. G. Hinks, J. D. Jorgensen, cond-mat/0104206 (11 Apr 2001). F. Bouquet et al., *Phys. Rev. Lett.* 87, 047001 (2001).
128. S. Tsuda et al. cond-mat/0104489 (25 Apr 2001) S. Tsuda, T. Yokoya, T. Kiss, Y. Takano, K. Togano, H. Kito, H. Ihara, and S. Shin *Phys. Rev. Lett. 87*, 177006 (2001).
129. F. Giubileo, D. Roditchev, W. Sacks, R. Lamy, and J. Klein cond-mat/0105146 (7 May 2001); *EuroPhys. Lett.* 58, 764 (2002)
130. F. Giubileo, cond-mat/0105592 (30 May 2001); F. Giubileo, D. Roditchev, W. Sacks, R. Lamy, D.X. Thanh, J. Klein, S. Miraglia, D. Fruchart, J. Marcus, and Ph. Monod, *Phys. Rev. Lett.* 87, 177008 (2001).

131. P. Szabo, et al. cond-mat/0105598 (30 May 2001). P. Szabo, P. Samuely, J. Kacmarck, T. Klein, J. Marcus, D. Fruchart, S. Miraglia, C. Marcenat, and A. G. M. Jansen, *Phys. Rev. Lett.* 87, 137005 (2001).
132. A. Bianconi, and A. Saccone, in *Studies of High Temperature Superconductors* vol. 38 Ed. A. V. Narlikar (NOVA Science NY 2001). Chapter 8.
133. F. Bouquet, Y. Wang, R.A. Fisher, D. G. Hinks, J. D. Jorgensen, A. Junod, N. E. Phillips, *Europhys. Lett.* 56, 856 (2001).
134. Y. Wang, T. Plackowski, and A. Junod, *Physica C* 355, 179 (2001).
135. H. Rosner et al. in *Studies of High Temperature Superconductors* vol. 38 Ed. A.V. Narlikar (NOVA Science NY 2001). Chapter 2.
136. R.A. Fisher et al. in *Studies of High Temperature Superconductors* Vol. 38 Ed. A. V. Narlikar Chapter 10 (NOVA Science NY 2001); cond-mat/0107072.
137. J.F. Annett, and S. Kruchinin *"New Trends in Superconductivity"* Kluwer Academic Publishers, Dodrecht, Netherlands (2002)
138. T. Haase, and K. Yamaji *J. Phys. Soc. Jpn.* 70, 2376 (2001).
139. A. Bianconi *Int. J. Mod. Phys. B* 16, 1591 (2002).
140. J.H. Jung, K.W. Kim, H.J. Lee, M.W. Kim, T.W. Noh, W.N. Kang, H.-J. Kim, E.-M. Choi, C. U. Jung, and S.-I. Lee *Phys. Rev. B* 65, 052413 (2002).
141. A.K. Pradhan, M. Tokunaga, Z.X. Shi, Y. Takano, K. Togano, H. Kito, H. Ihara, and T. Tamegai, *Phys. Rev. B* 65, 144513 (2002).
142. A.B. Kuzmenko, F.P Mena, H.J.A. Molegraaf, D. van der Marel, B. Gorshunov, M. Dressel, I.I. Mazin, J. Kortus, O.V. Dolgov, T. Muranaka, J. Akimitsu, *Solid State Commun.* 121, 479 (2002).
143. A. Bianconi, S. Agrestini, D. Di Castro, G. Campi, G. Zangari, N.L. Saini, A. Saccone, S. De Negri, M. Giovannini, G. Profeta, A. Continenza, G. Satta, S. Massidda, A. Cassetta, A. Pifferi and M. Colapietro, *Phys. Rev. B* 65, 174515 (2002).
144. O. De La Pena, A. Aguayo, and R. de Coss, *Phys. Rev. B* 66, 012511 (2002).
145. A. Brinkman, A.A. Golubov, H. Rogalla, .V. O Dolgov, J. Kortus, Y. Kong , O. Jepsen, and O.K. Andersen, *Phys. Rev. B* 65, 180517 (2002).
146. A.A. Golubov, J. Kortus, O.V. Dolgov, O. Jepsen, Y. Kong, O.K. Andersen, B.J. Gibson, K. Ahn, and R.K. Kremer, *J. Phys.: Condens. Matter* 14, 1353 (2002).
147. I.I. Mazin, Andersen, O. Jepsen, O.V. Dolgov, J. Kortus, A.A. Golubov, A.B. Kuz'menko, D. van der Marel, *Phys. Rev. Lett* 89, 107002 (2002).
148. H.J. Choi, D. Roundy, H. Sun, M.L. Cohen, and S.G. Louie, *Phys. Rev. B* 66, 020513 (2002).
149. H.J. Choi, D. Roundy, H. Sun, M.L. Cohen, and S.G. Louie, *Nature (London).* 418, 758 (2002).
150. A. Brinkman, A.A. Golubov, H. Rogalla, O.V. Dolgov, J. Kortus, Y. Kong, O. Jepsen, O.K. Andersen, *Phys. Rev. B* 65, 180517 (2002).
151. H. Schmidt, J.F. Zasadzinski, K.E. Gray, and D.G. Hinks, *Phys. Rev. Lett.* 88, 127002 (2002).
152. V.G. Kogan, *Phys. Rev. B* 66, 020509 (2002).

153. H. Rosner, J. M. An, W. E. Pickett, S.-L. Drechsler, *Phys. Rev. B* 66, 024521 (2002).
154. R.S. Gonnelli, D. Daghero, G.A. Ummarino, V.A. Stepanov, J. Jun, S.M. Kazakov, and J. Karpinski, *Phys. Rev. Lett.* 89, 247004 (2002).
155. C.P. Moca, and C. Horea, *Phys. Rev. B* 66, 052501 (2002).
156. M.-S. Kim, J.A. Skinta, T.R. Lemberger, W.N. Kang, H.-J. Kim, E.-M. Choi, and S.-I. Lee, *Phys. Rev. B* 66, 064511 (2002).
157. Z.-Z. Li, H.-J. Tao, Y. Xuan, Z.-A. Ren, G.-C. Che, and B.-R. Zhao, *Phys. Rev. B* 66, 064513 (2002).
158. M. Iavarone, G. Karapetrov, A.E. Koshelev, W.K. Kwok, G.W. Crabtree, D.G. Hinks, W. N. Kang, E.-M. Choi, H. J. Kim, H.-J. Kim, and S.I. Lee, *Phys. Rev. Lett.* 89, 187002 (2002).
159. F. Bouquet, Y. Wang, I. Sheikin, T. Plackowski, A. Junod, S. Lee, and S. Tajima, *Phys. Rev. Lett.* 89, 257001 (2002).
160. H. Uchiyama, K.M. Shen, S. Lee, A. Damascelli, D.H. Lu, D.L. Feng, Z-X. Shen, and S. Tajima, *Phys Rev Lett* 88, 15700 (2002).
161. H. Nagao, S.P. Kruchinin, A.M. Yaremko, and K. Yamaguchi *Int. J. Mod. Phys. B,* 16, 23, 3419 (2002).
162. A. Bussmann-Holder and A. Bianconi, *Phys. Rev.* B 67, 132509 (2003).
163. I.I. Mazin, and V.P. Antropov, *Physica C* 385, 49 (2003).
164. M. Schechter, J. von Delft, Y. Imry, and Y. Levinson, *Phys. Rev. B* 67, 064506 (2003).
165. J.J. Rodriguez-Nunes and A.A. Schmidt, *Phys. Rev. B* 68, 224512 (2003)
166. P. Samuely, Z. Holanov, P. Szabu, J. Kacmarcìk, R.A. Ribeiro, S.L. Bud'ko, and P.C. Canfield, *Phys. Rev. B* 68, 020505 (2003).
167. A. Shibata, M. Matsumoto, K. Izawa, Y. Matsuda, S. Lee, and S. Tajima, *Phys. Rev. B 68*, 060501 (2003).
168. R.A. Ribeiro, S.L. Bud'ko, C. Petrovic, and P.C. Canfield, *Physica C* 384, 227 (2003).
169. H. Schmidt, K.E. Gray, D.G. Hinks, J.F. Zasadzinski, M. Avdeev. J.D. Jorgensen and C.J. Burley, *Phys. Rev. B* 68, 060508 (2003).
170. M. Putti, M. Affronte, P. Manfrinetti, and A. Palenzona, *Phys. Rev. B* 68, 094514 (2003).
171. K.A. Yates, Burnell, N.A. Stelmashenko, D.-J. Kang, H.N.Lee, B. Oh, and M.G. Blamire, *Phys. Rev. B* 68, 220512 (2003).
172. T. Dahm and N. Schopohl, *Phys. Rev. Lett.* 91, 017001 (2003).
173. S. Tsuda, T. Yokoya, Y. Takano, H. Kito, A. Matsushita, F. Yin, J. Itoh, H. Harima, and S. Shin, *Phys. Rev. Lett.* 91, 127001 (2003).
174. Y. Xu, M. Khafizov, L. Satrapinsky, P. K˙s, A. Plecenik, and R. Sobolewski, *Phys. Rev. Lett.* 91, 197004 (2003).
175. S. Souma, Y. Machida, T. Sato, T. Takahashi H, Matsui S-C, Wang H, Ding A, Kaminski J.C, Campuzano S, Sasaki and K, Kadowaki, *Nature (London),* 423, 65 (2003).

176. A. Bianconi, and S. Agrestini, "*$Sc_{1-x}Mg_xB_2$ tuned at a shape resonance for T_c amplification*", European Patent No. 14477857, European Patent Bull. 2004/34 (18.08.2004).
177. S. Agrestini, C. Metallo, M. Filippi, L. Simonelli, G. Campi, C. Sanipoli, E. Liarokapis, S. De Negri, M. Giovannini, A. Saccone, A. Latini, and A. Bianconi, *Phys. Rev. B* 70, 134514 (2004).
178. Y. Bugoslavsky, Y. Miyoshi, G.K. Perkins, A.D. Caplin, L.F. Cohen, A.V. Pogrebnyakov, and X. X. Xi *Phys. Rev. B* 69, 132508 (2004).
179. B. Kang, H.-J. Kim, M.-S. Park, K.-H. Kim, and S.-I. Lee, *Phys. Rev. B* 69, 144514 (2004).
180. G.A. Ummarino, R.S.Gonnelli, S. Massidda and A.Bianconi, *Physica C* 407, 121 (2004).
181. A. Brinkman, A.A. Golubov and M.Yu. Kupriyanov, *Phys. Rev. B* 69, 214407 (2004).
182. S. Tsuda, T. Yokoya, S. Shin, M. Imai, and I. Hase, *Phys. Rev. B* 69, 100506 (2004).
183. S. Tsuda et al. cond-mat/0409219.
184. S. Shin, T. Kiss, S. Tsuda, T. Yokoya, and C. Chen, T. Togashi, S. Watanabe, S. Lee, H. Uchiyama, and S. Tajima, *Int, Conf."Stripes 04" Rome 27 Sept-1 Oct 2004 Book of Abstracts unpublished.*
185. G. Binnig, A. Baratoff, H.E. Hoenig and J. G. Bednorz, *Phys. Rev. Lett.* 45, 1352 (1980).
186. D.F. Agterberg, T.M. Rice, and M. Sigrist, *Phys. Rev. Lett.* 78, 3374 (1997).
187. D.F. Agterberg, *Phys. Rev. B* 60, 749 (1999)
188. M.E. Zhitomirsky, and T. M. Rice, *Phys. Rev. Lett.* 87, 057001 (2001).
189. S.V. Shulga, S.L. Drechsler, G. Fuchs, K.H. Muller, K. Winzer, M. Heinecke and K. Krug, *Phys. Rev. Lett.* 80, 1730 (1998).
190. E. Boaknin, M.A. Tanatar, J. Paglione, D. Hawthorn, F. Ronning, R.W. Hill, M. Sutherland, L. Taillefer, J. Sonier, S.M. Hayden, and J.W. Brill, *Phys. Rev. Lett.* 90, 117003 (2003).
191. I.M. Lifshitz, *Soviet Physics JEPT* 11, 1130 (1960).
192. A. Bianconi, in "*X-ray and Inner Shell Processes*" AIP Conf. Proc. (2002) pp. 13-18
193. E. Tiesinga, B.J. Verhaar, and H.T.C. Stoof, *Phys. Rev. A* 47, 4114 (1993)
194. S. Inouye et al., *Nature* 392, 151 (1998)
195. C.A. Regal, M. Greiner, and D.S. Jin, *Phys. Rev. Lett.* 92, 040403 (2004)
196. Y. Guo, Y.-F. Zhang, X.-Yu Bao, T.Z, Han, Z. Tang, L.-X Zhang, W.-G. Zu, E.G. Wang, Q. Niu, Z, Q, Qiu, Y.-F Jia, Z.-X. Zhao, and Q.K. Xue, *Science* 306, 1915 (2004).
197. B. Aronsson, T. Lundstrom, and S. Rundquist, *Borides, Silicides and Phosphides* Methuem, London (1965).
198. J.L. Hoard, and R.E. Hughes, *The Chemistry of Boron and its Compounds* J Wiley, NY (1967).
199. W. Hoffmann, and W. Jänicke, *Naturwiss* 23, 851 (1935).

200. V. Russel, R. Hirst, F. Kanda and A. King, *Acta Cryst.* 6, 870 (1953).
201. E. Jones and B. Marsh *J Am. Chem. Soc.* 76, 1434 (1954).
202. A.F. Goncharov, and V.V. Struzhkin, *Physica C* 385, 117 (2003).
203. H. Olijnyk, A.P. Jephcoat, D.L. Novikov, N.E. Christensen, *Phys. Rev. B* 62, 5508 (2000).
204. R. de Coss et al. Presented at Int, Conf."Stripes 04" Rome 2004, unpublished.
205. G. G. N. Angilella, A. Bianconi, and R. Pucci cond-mat/ 0503544 (22 March 2005)
206. N. Kristoffel T. Örd, and K. Rago, *Europhysics Lett.* 61, 109 (2003); N. Kristoffel, and P. Rubin in "*Simmetry and Heterogeneity in High Temperature Superconductivity*" A. Bianconi Ed. Kluwer Academic Publisher Dordrecht (2005).
207. S.P. Kruchinin, and H. Nagao in "*Simmetry and Heterogeneity in High Temperature Superconductivity*" A. Bianconi Ed. Kluwer Academic Publisher Dordrecht (2005)

I.3

MODELLING CUPRATE GAPS IN A COMPOSITE TWO-BAND MODEL

N. Kristoffel and P. Rubin
Institute of Physics, University of Tartu, Riia 142, 51014 Tartu, Estonia

Abstract: A simple model to cover the two-component scenario of cuprate superconductivity is developed. Interband pairing interaction acts between itinerant and defect states created by doping. Two defect system subbands correspond to "hot" and "cold" regions of the momentum space. Superconductivity energetic characteristics vs doping are compared to experimental findings. Transformations of two pseudogaps into superconducting and normal state gaps can be traced. Doping concentrations where the band components begin to overlap determine essential borders on the phase diagram. Qualitative agreement with observations is present including the effect of photodoping.

Key words: Cuprate; Two-band Model; Gaps

1. INTRODUCTION

Pair transfer interaction between the states of an electron system components can cause the gauge symmetry breaking realized in superconductivity. This circumstance forms the basis of the two-band model of superconductivity known already during a considerable time [1,2]. The basic advantage of such approaches consists in the possibility to reach pairing by a repulsive interband interaction which operates in a considerable volume of the momentum space. An electronic energy scale is

A. Bianconi (ed.), Symmetry and Heterogeneity in High Temperature Superconductors, 55–64.

correspondingly introduced into determination of the transition temperature. Further elaborations have been remained without distinct bases of applications until the high-T_c cuprate superconductivity has been discovered. Under numerous approaches looking for the basic mechanism of this event the two-band approaches have been used as one of the possible channels. The nature of the electron band components used has been variable (for review see e.g. [3-5]).

The progress in experimental investigations has revealed some coupled general properties of cuprates which are connected with the necessary doping treatment. First, the structure of doped cuprates has been found to be essentially inhomogeneous on the nanoscale (stripes, tweed patterns, granularity) with the associated electronic phase separation [6-12] in the CuO_2 planes. At the same time it has been found that the hole doping creates a new band of defect-states in the cuprate charge-transfer gap [13-24]. Various data have indicated the functioning of itinerant and "defect"-type carriers in the basic physics of cuprate superconductivity. A corresponding "two-component" scenario [25-27] has been formulated. The hole-poor material can be considered as remaining the source for the itinerant band and the part of the distorted material bearing doped holes as creating the defect-band. The application of two-band models in the frame of the two-component scenario becomes natural. Such kind of approaches are strongly supported by the recent tunneling and angle resolved photoemission spectroscopy data [28,29], which have revealed the two-gap behaviour of cuprates. As a minimum it means the presence of a pseudogap plus a superconductivity gap [30], or two pseudogaps and a superconductivity gap [31-35].

However, meantime the new superconductor MgB_2 [36] has been discovered, which shows a clear two-gap behaviour [37]. Two-band models have been applied to MgB_2 [38-39] and at present this system is addressed as the reigning representative of the two-band superconductivity.

Concerning the cuprates it is now clear that a nonrigid basic electron spectrum must be used. Doping creates not only carriers but prepares the whole background (incorporating also structural changes) for the appearance of superconductivity. Under corresponding approaches there are ones which take as the basis a narrow defect (bipolaron) band above the itinerant band [23,40-48], or stripe-induced minibands [11,42].

It seems that a self-consistent scheme of cuprate multigap properties over the phase diagram incorporating the essence of numerous experimental investigations is yet missing. The statements about the relations and coexistence of various gaps for different doping regions and energy scales are partly controversial. In such a situation elaboration of a simple and easily tractable model for the calculation of cuprate energetic characteristics with the following projection of results on the experimental data can make sense. Correspondingly, in [45,46] a model for the evolution of the cuprate electron spectrum under hole doping has been postulated. The central result of the model has been the natural explanation of the underdoped state pseudogap by the narrowing gap between the defect and itinerant band states. This model has been extended by the account of interband vibronic interaction [46,47]. It allowed to extend the pseudogap behaviour behind the optimal doping as observed [48].

Structurally the defect subsystem is anisotropic and it is manifested in differences of gap characteristics measured for various regions of the momentum space [18,28,29,49]. The well-expressed pseudogap is connected with the neighbourhood of the "hot" $(\pi, 0)$-type points in the Brillouin zone. The spectrum at (π, π)-type "cold" points seems to be nongapped, as in the *d*-wave concept [50].

In what follows we present a generalization of [45,46] by introducing two subbands to distinguish the role of "hot" and "cold" regions of the defect-system spectrum, cf. [43].

2. THE MODEL

The itinerant (valence) band (γ) is taken to be lying between energies $\xi = -D$ and $\xi = 0$ with the number of states normalized to $1 - c$. Here c characterizes the doped hole concentration and must be scaled for a given case. The defect bands (α; β) created by doping occupy the energy intervals $d_1 - \alpha c$ and $d_2 - \beta c$ correspondingly with the weight of states $c/2$. The 2*D* densities of states read $\rho_\alpha = (2\alpha)^{-1}$; $\rho_\beta = (2\beta)^{-1}$; $\rho_\gamma = (1-c)D^{-1}$. By doping the β and γ-band join at $c_\beta = d_2\beta^{-1}$ and a common overlapping energy distribution of all the bands is reached at $c_\alpha = d_2\alpha^{-1}$. The α-band is attributed to the "hot" region. At $c < c_\beta$ the chemical potential $\mu = d_2 - \beta c$ is connected with the "cold" β band. For $c > c_\beta$, $\mu = (d_2 - \beta c)\,[1 + 2\beta(1 - c)D^{-1}]^{-1}$ intersects both (β; γ) bands. Further for $c \geq c_0$ ($d_1 - \alpha c_0 = \mu$) one finds $\mu = [\alpha d_2 + \beta d_1 - 2\alpha\beta c][\alpha + \beta + (1-c)2\alpha\beta D^{-1}]^{-1}$ which intersects all the three overlapping bands.

Superconductivity characteristics will be calculated supposing the interband transfer of pairs formed from the particles of the same band according to the mean-field Hamiltonian

$$H = \sum_{\sigma\vec{k}s} \in_\sigma (\vec{k}) \alpha^+_{\sigma\vec{k}s} \alpha_{\sigma\vec{k}s} + \sum_{\vec{k}} \Delta_\gamma(\vec{k})[\alpha_{\gamma\vec{k}\uparrow}\alpha_{\gamma-\vec{k}\downarrow} + \alpha^+_{\gamma-\vec{k}\downarrow}\alpha^+_{\gamma\vec{k}\uparrow}] - \sum_{\vec{k},\sigma'} \Delta_{\sigma'}(\vec{k})[\alpha_{\sigma'\vec{k}\uparrow}\alpha_{\sigma'-\vec{k}\downarrow} + \alpha^+_{\sigma'-\vec{k}\downarrow}\alpha^+_{\sigma'\vec{k}\uparrow}] \qquad (1)$$

Here $\epsilon_\sigma = \xi_\sigma - \mu, \sigma = \alpha, \beta, \gamma$, $\sigma' = \alpha, \beta$, usual designation for spins (s) and electron operators apply. The superconductivity order parameters are defined as

$$\begin{aligned} \Delta_\gamma(\vec{q}) &= 2\sum_{\vec{k},\sigma'} W(\vec{q},\vec{k})\langle \alpha_{\sigma'\vec{k}\uparrow}\alpha_{\sigma'-\vec{k}\downarrow}\rangle \\ \Delta_{\sigma'}(\vec{q}) &= 2\sum_{\vec{k}} W(\vec{q},\vec{k})\langle \alpha_{\gamma-\vec{k}\downarrow}\alpha_{\gamma\vec{k}\uparrow}\rangle \end{aligned} \qquad (2)$$

The interband interaction constant $W > 0$ corresponds to repulsion with possible electronic and electron-phonon contributions. The quasiparticle energies read $E_\sigma(\vec{k}) = \sqrt{\in^2_\sigma(\vec{k}) + \Delta^2_\sigma(\vec{k})}$. In what follows the momentum dependences will be accounted for only by the α–β differentation at a constant W (different energy intervals for (α,β) in gap-equation integrals). It is taken $\Delta_\alpha = \Delta_\beta$. At T_c, $\Delta_{\alpha,\gamma} \rightarrow 0$ simultaneously. For $W > 0$ two s-type order parameters appear with opposite signs [3]; expr. (1) uses positive $\Delta_{\alpha,\gamma}$.

3. THE GAPS AND RELATIONS BETWEEN THEM

Now we look on the minimal excitation energies of the quasiparticles which reflect the presence of gaps in the charge-excitation-channel of the superconductor.

In the low underdoping region for $c < c_\beta$

$$\begin{aligned} E_\alpha(min) &= \Delta_l = \sqrt{(d_1 - \alpha c - \mu)^2 + \Delta^2_\alpha} \\ E_\beta(min) &= \Delta_\alpha \\ E_\gamma(min) &= \Delta_s = \sqrt{\mu^2 + \Delta^2_\gamma} \end{aligned} \qquad (3)$$

It is seen that in the normal state Δ_l and Δ_s survive and can be interpreted as the large and small pseudogap. Passing on to the optimal doping ($c \geq c_\beta$, $c < c_0$) the small pseudogap transforms into the itinerant superconducting gap. For $c \geq_s c_0$ the large pseudogap becomes also quenched and is smoothly transformed to the defect system ($\Delta_\alpha < \Delta_\gamma$) superconducting gap.

Concerning the manifestation of various gaps of the model, the E_γ excitations become attributed to the "hot" spectrum. The "cold" spectrum is usually considered as nongapped [28,49]. If one takes account of this by multiplying the defect subsystem Δ_α by a d-wave symmetry factor, the cold spectrum becomes empty. At $c < c_\beta$ the appearance of two pseudogaps is expected. In the basic optimal doping region ($c \geq c_\beta$, $c < c_0$) the spectrum involves the large pseudogap and the superconducting gap Δ_γ. At heavier overdoping $c > c_0$ the spectrum is expected to contain two superconductivity gaps – the defect Δ_α manifested by additional spectral weight inside of Δ_γ. Such kind of the phase diagram is in qualitative agreement with the results found experimentally.

In spite of that the majority of gap manifestations appears in the "hot" spectrum, the "cold" electrons act essentially in building up superconductivity. Beyond c_β the cold subband acts as the necessary overlapping partner to achieve a high T_c by the interband mechanism. The presence of pseudogaps in the present model is reduced to the creation of a multicomponent electron liquid by doping.

4. ILLUSTRATION AND DISCUSSION

The illustrative calculation of cuprate energetic characteristics on the doping scale (Fig.1) has been made with a plausible set of parameters: $D = 2$; $d_1 = 0.3$; $d_2 = 0.1$; $\alpha = 0.66$; $\beta = 0.33$ and $W = 0.28$(eV). The maximal $T_c = 125$ K is reached for $c = 0.57$ and $c_\alpha = 0.45$; $c_\beta = 0.30$; $c_0 = 0.57$. A scaling for a typical cuprate hole doping has been made through $p = 0.4\ c$. The large pseudogap Δ_l is extended until the moderate overdoping in agreement with observations [17,50-52]. Further it is smoothly transformed into the defect system superconducting gap. This agrees with the experimental result [53]. The manifestation of both superconductivity gaps expected at $c > c_0$ (of two Fermi surfaces) is often debated. However, the common Fermi surface is then intersected by the band components at different wave vectors and the large Δ_γ can remain masked. At intermediate dopings the large pseudogap and the superconducting gap appear and cross close to the optimal doping, as found experimentally [52]. Δ_l and Δ_γ are connected with different but non competing order parameters of defect and itinerant subsystems. Spectrally Δ_l must be related to the hump-feature. Eventually it remains preserved for $T > T_c$ (insert in Fig.1) at the dopings where T_c is optimized [52] and it is shifted to larger energies with reduced doping [54].

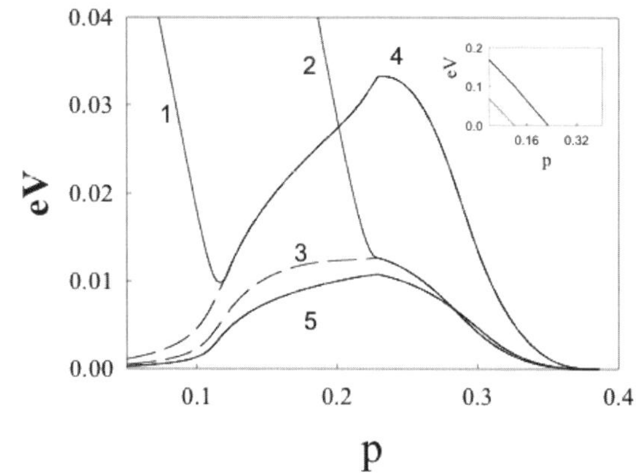

Figure I:3:1. Energetic characteristics of a model cuprate on the doping scale. 1 – the underdoped state small pseudogap Δ_s; 2 – the large pseudogap Δ_l; 3 – the defect system superconducting gap Δ_α; 4 – the itinerant system superconducting gap Δ_γ 5 – T_c. The insert shows normal state gaps. $p_\alpha = 0.18$; $p_\beta = 0.12$ $p_0 = 0.23$; $p(T_{cm}) = 0.23$]

The common manifestation of two underdoped state pseudogaps is expected theoretically. This has been recently established experimentally for the La and Bi-cuprates [31-35]. The small pseudogap is known to develop smoothly from the larger superconducting gap [18,31,32]. That is comparable with the transformation $\Delta_s \leftrightarrow \Delta_\gamma$ at c_β.

In Fig.2 the calculated Uemura type plot [22] is given. It is considered as universally characteristic for cuprates.

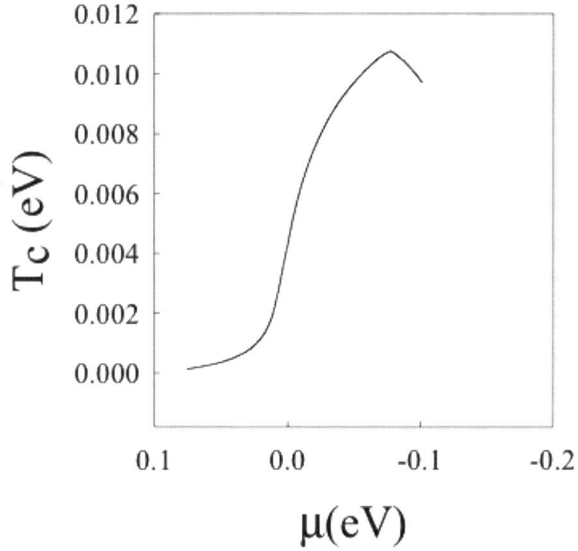

Figure I:3:2. Transition temperature vs. chemical potential for a model cuprate (an Uemura type plot)

The nonrigid dynamics of the electron spectrum with doping as characteristic for the present model points to special concentrations where the band components begin to overlap. A special critical doping region on the cuprate phase diagram where the properties of the electron liquid change essentially is well known [15,16,55]. This is of utmost importance in quantum critical point scenarios [15,42]. The (large) pseudogap is lost when passing this c_k border. It is tempting to relate c_k to c_0. Then the progressive doping is expected to restart the Fermi-liquid behaviour (cf. [15,16]) by building up a mixed electronic system of overlapping defect and itinerant bands. It appears when the bottom edge of the hot defect states lowers so that μ intersects all three bands. The Fermi surface becomes more and more electron-like and the difference between the itinerant and defect subsystem is washed up. Experiments on the normal state [55] show that a quantum metal-insulator transition appears at c_k. For smaller dopings the hot quasiparticles become insulating whereas the cold quasiparticles remain metallic. This is the same story which follows from the present model.

In the last paragraph we demonstrate that the present model is able to reproduce also the experimental findings on the photodoping effect.

5. PHOTODOPING EFFECT

Investigations of the photodoping effect in $YBa_2Cu_3O_{6+x}$ [56,57] and $Tl_2Ba_2CaCu_2O_8$ [58] have revealed an increase of T_c. On the contrary, for $Bi_2Sr_2CaCu_2O_{8+\delta}$ a photoinduced depression of T_c has been found [59]. These cases differ seemingly by the localization of the photoelectrons. In the case of Y-123 the photoelectrons become localized out of the CuO_2 planes. In the present model it is characterized by the simple substitution $c \to c + x$. It means that the sign of the photodoping effect is determined by dT_c/dc. For $YBa_2Cu_3O_{6+\delta}$ it gives the enhancement of T_c for compositions from $\delta = 0.4$ to $\delta = 1$. The experimental $\Delta T_c(\delta)$ curve of [57] is roughly reproduced by the left part of the theoretical curve in Fig.3 with $c \sim \delta/2$.

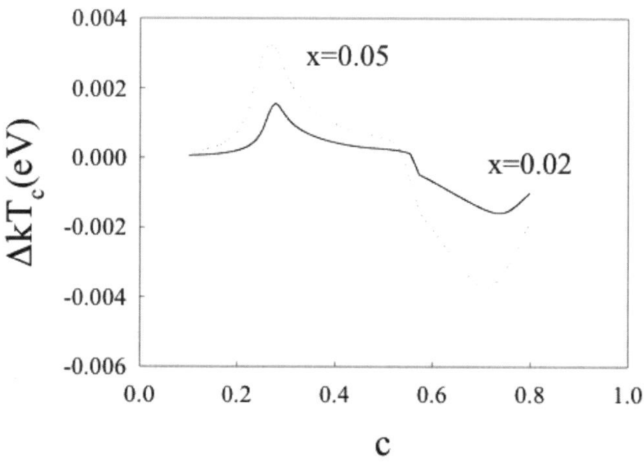

Figure I:3:3. The photoinduced shift of T_c for the case of photoelectrons localized out of the CuO_2 planes

6. CONCLUSION

The present simple and partly postulative model seems to be able to reproduce the observed behaviour of the cuprate energetic characteristics. The possible coexistence of pseudogap with the superconducting gap of the other partner subsystem has been demonstrated. The transformation of the pseudogap into the superconducting gap of the own subsystem by extended doping can appear. The pseudogap and the normal state gap are connected by the quenching of the superconducting gap contribution. These relations are reduced to the multicomponent nature of the cuprate electron spectrum reorganized by doping.

The results following from the present model (it must not be taken too literally) seem to point to the plausible nature of its basic content. A wide freedom remains to its improvement in various aspects.

This work was supported by the Estonian Science Foundation Grant No 4961.

REFERENCES

1. H. Suhl, B.T. Matthias, and L.R. Walker, Phys. Rev. Lett. **3**, 552, (1959).
2. V.A. Moskalenko, *Electromagnetic and kinetic properties of superconducting alloys with overlapping electron bands*, Shtiinza, Kishinev (1976) and references therein (1976).
3. N. Kristoffel, P. Konsin, and T. Örd, *Nuovo Cimento* **17**, 1 (1994).
4. V.A Moskalenko, M.E. Palistrant, and V.M. Vakalyuk, *Uspekhi Fiz. Nauk.* **161**, 155 (1991).
5. N. Plakida, *High-temperature superconductivity*, Springer, Berlin (1995).
6. V.J. Emery, and S.A. Kivelson, *Physica C* **209**, 597 (1995).
7. M.J. Salkola, et al., *J. Supercond.* **9**, 401 (1996).
8. T. Egami, *J. Low Temp.* **105**, 791 (1996).
9. A. Bianconi, et al. *Phys. Rev. B* **54**, 12018 (1996).
10. J.M. Tranquada, J. *Supercond.* **9**, 397 (1997).
11. A. Bianconi et al., *Physica C* **296**, 269 (1998).
12. J.C. Phillips, and J. Jung, *Phil. Mag. B* **81**, 745 (2001).
13. A.V. Chubukov, and D.K. Morr, *Phys. Repts.* **288**, 355 (1997).
14. M.G. Zacher, et al., *Phys. Rev. B* **85**, 2585 (2000).
15. D. Di Castro, et al. *Eur. Phys. J. B* **18**, 617 (2000).
16. P. Schwaller, et al., *Eur. Phys. J. B* **18**, 215 (2000).
17. E. Dagotto, *Rev. Mod. Phys.* **66**, 763 (1994).
18. A. Ino, et al., *Phys. Rev. B* **65**, 094504 (2002).
19. Y. Ando, *Phys. Rev. Lett.* **87**, 017001 (2001).
20. C.C. Homes, et al., *Phys. Rev. B* **67**, 184516 (2003).
21. H. Romberg, et al. *Phys. Rev. B* **42**, 8768 (1990).
22. Y.J. Uemura, in *"Polarons and Bipolarons"* Eds. E.K.H. Salje, A.S. Alexandrov and W.Y. Liang, Cambridge Univ., p. 453 (1995).
23. A.S. Alexandrov, and P.P. Edwards, *Physica C* **331**, 97, (2000).
24. J. Lorenzana, and G. Seibold, *Phys. Rev. Lett.* **89**, 136401 (2002).

25. D. Mihailovic, and K.A. Müller, *in Materials Aspects of High-T_c Superconductivity, NATO ASI,* Kluwer, Dordrecht, 1 (1997).
26. K.A. Müller, *Physica C* **341-348**, 11 (2000).
27. A. Bianconi and N.L. Saini, *Stripes and Related Phenomena,* Kluwer Acad. Publ., N-Y, 2000.
28. T. Timusk, and B. Statt, *Rep. Progr. Phys.* **62**, 61 (1999).
29. A. Damascelli, Z. Hussain, and Z.-X. Shen, *Rev. Mod. Phys.* **75**, 473 (2003).
30. A. Mouraschkine, *Physica C* **341-348**, 917 (2000).
31. R.M. Dipasupil, et al., *J. Phys. Soc.* Jpn. **71**, 1535 (2002).
32. T. Sato, et al., *Physica C* **341-348**, 815 (2000).
33. A. Fujimori, et al., *Physica C* **341-348**, 2067 (2000).
34. T. Takahashi, et al., *J. Phys. Chem. Solids* **62**, 41, (2001).
35. M. Oda, N. Momono, and M. Ido, in *New Trends in Superconductivity*, J.F.Annett, S.Kruchinin Eds., Kluwer Acad. Publ., Dordrecht, 177, (2002).
36. P.C., Canfield, S.L. Bud'ko, and D.K. Finnmore, *Physica C* **385**, 1 (2003). A. Bianconi, et al. , *J. Phys.: Condens. Matter* 13, 7383 (2001); A. Bianconi, and M. Filippi, chapter 2 of this volume.
37. A.Y. Liu, I.I. Mazin, and J. Kortus, *Phys. Rev. Lett.* **87**, 087005 (2001).
38. A.A. Golubov, ct al., *Phys. Rev. B* **66**, 054524 (2002).
39. N. Kristoffel, T. Örd, and K. Rägo, *Europhys. Lett.* **61**, 109 (2003).
40. N. Kristoffel, *Phys. Stat. Sol. B* **210**, 195, (1998).
41. L.P. Gor'kov, and A.V. Sokol, *Pis'ma ZETF* **46**, 333 (1987).
42. A. Bussmann-Holder et al., *J.Phys.: Cond. Matter* **13**, L169 (2001).
43. A. Perali, et al., Phys. Rev. B **62**, R9295 (2000).
44. R. Micnas, S. Robaszkiewicz, and A. Bussmann-Holder, *Physica C* **387**, 58 (2003).
45. N. Kristoffel, and P. Rubin, *Physica C* **356**, 171 (2001).
46. N. Kristoffel, and P. Rubin, *Solid State Commun.* **122**, 265 (2002).
47. N. Kristoffel, and P. Rubin, *Eur. Phys. J. B* **30**, 495 (2002).
48. N. Kristoffel, Modern *Phys. Lett. B* **17**, 451 (2003).
49. C.C. Tsuei, and J. Kirtley, *Rev. Mod. Phys.* **72**, 969 (2000).
50. H. Ding, et al., *Nature* **382**, 51 (1996).
51. M. Moraghebi, et al., *Phys. Rev. B* **63**, 214513 (2001).
52. V.M. Krasnov, et al., *Phys. Rev. Lett.* **84**, 5860 (2000).
53. D. Mihailovic, et al., *Physica C* **341-348**, 1731 (2000).
54. N. Miyakawa, et al., *Phys. Rev. Lett.* **83**, 1018 (1999).
55. F. Venturini, et al., *Phys. Rev. Lett.* **89**, 107003 (2002).
56. V.I. Kudinov, *Physica B* **194-196**, 1963 (1994).
57. E. Osquiquil, et al., *Phys. Rev. B* **49**, 3675 (1994).
58. H. Szymczak, et al., Europhys. Lett. **35**, 451 (1996).
59. K. Tanabe, et al. Phys. Rev. B **52**, R13152 (1995).

I.4

MULTI-GAP SUPERCONDUCTIVITY IN MgB_2

S. P. Kruchinin[1], and H. Nagao[2]
[1]*Bogolyubov Institute for Theoretical Physics, The Ukrainian National Academy of Science, Kiev 252143, Ukraine*
[2]*Department of Computational Science, Faculty of Science, Kanazawa University Kakuma, Kanazawa 920-1192, Japan*

Abstract: We using our two-band model for the explanation of two coupled superconductivity gaps for MgB_2.To study the effect of the increasing of T_c in MgB_2 due to the enhanced interband pairing scattering. We have proposed two channel scenario of superconductivity: the conventional channel which is connected with BCS mechanism in different zone and the unconventional channel which describes the transfer or tunneling of Cooper pair between two bands.

Key words: Multi-gap, Two-band model, Interband Pairing.

1. INTRODUCTION

Recent discovery of superconductivity of MgB_2 with T_c=39K, which is the highest as for intermetallics [1], has been recognized a multi-gap superconductor[2]. Although multi-gap superconductivity had been discussed theoretically [3-5] in the 1958, the multi-gap superconductivity has been observed experimentally [6] in the 1980s. MgB_2 is the first material in which its effects are so dominant and its implications so thoroughly explored. Nature has let us glimpse a few or her multi-gap mysteries and has challenged us. Recent band calculations [7,8] of MgB_2 cite with the

A. Bianconi (ed.), Symmetry and Heterogeneity in High Temperature Superconductors, 65–75.

McMillan formula of transition temperature have supported the e-p interaction mechanism for the superconductivity. In this superconductivity, the possibility of two-band superconductivity has also been discussed in relation to two gap functions experimentally and theoretically. Very recently, two-band or multi-band superconductivity has been theoretically investigated in relation to superconductivity arising from coulomb repulsive interactions. Two-band model was first introduced by Kondo [5]. We have also investigated anomalous phases in two-band model by using the Green function techniques[10-15]. Recently, we have pointed out the importance of many-band effects in high-T_c superconductivity [10-14]. The expressions of the transition temperature for several phases have been derived, and this approach has been applied to superconductivity in molecular crystals by charge injection and field-induced superconductivity [11]. In the previous papers [10-12], we have investigated superconductivity by using the two-band model and a two-particle Green function techniques. We have applied the model to an electron-phonon mechanism for the traditional BCS method, an electron-electron interaction mechanism for high- T_c superconductivity [9], and a cooperative mechanism. In the framework of the two-particle Green function techniques [16], it has been shown that the temperature dependence of the superconductivity gap for high- T_c superconductors is more complicated than predicted in the BCS approach. In paper [8] we have investigated the phase diagrams for two-band model superconductivity, using the renormalization group approach. We have discussed the possibility of a cooperative mechanism in the two-band superconductivity in relation to high-T_c superconductivity and to study the effect of the increasing of T_c in MgB_2 due to the enhanced interband pairing scattering.

In this paper, we investigate our two-band model for the explanation the multi-gap superconductivity of MgB_2. We apply the model to an electron-phonon mechanism for the traditional BCS method, an electron-electron interaction mechanism for high- T_c superconductivity, and a cooperative mechanism in relation to multi-band superconductivity.

2. THEORETICAL BACKGROUND

In this section, we briefly summarize a two-band model for superconductivity.

2.1 Hamiltonian

We start from the Hamiltonian for two-bands *i* and *j* ;

$$H = H_0 + H_{\rm int}\,, \tag{1}$$

where

$$H_0 = \sum_{\mathbf{k},\sigma} \Big[[\varepsilon_i - \mu] a^{+}_{i\mathbf{k}\sigma} a_{i\mathbf{k}\sigma} + [\varepsilon_j - \mu] a^{+}_{j\mathbf{k}\sigma} a_{j\mathbf{k}\sigma} \Big], \tag{2}$$

$$\begin{aligned} H_{\rm int} = \frac{1}{4} \sum_{\delta(\mathbf{p}_1+\mathbf{p}_2,\mathbf{p}_3+\mathbf{p}_4)} \sum_{\alpha\beta\gamma\delta} \Big[& \Gamma^{iiii}_{\alpha\beta\gamma\delta} a^{+}_{i\mathbf{p}_1\alpha} a^{+}_{i\mathbf{p}_2\beta} a_{i\mathbf{p}_3\gamma} a_{i\mathbf{p}_4\delta} + (i \to j) \\ & + \Gamma^{iijj}_{\alpha\beta\gamma\delta} a^{+}_{i\mathbf{p}_1\alpha} a^{+}_{i\mathbf{p}_2\beta} a_{j\mathbf{p}_3\gamma} a_{j\mathbf{p}_4\delta} + (i \to j) \\ & + \Gamma^{ijij}_{\alpha\beta\gamma\delta} a^{+}_{i\mathbf{p}_1\alpha} a^{+}_{j\mathbf{p}_2\beta} a_{i\mathbf{p}_3\gamma} a_{j\mathbf{p}_4\delta} + (i \to j) \Big], \end{aligned} \tag{3}$$

Γ is the bare vertex part:

$$\Gamma^{ijkl}_{\alpha\beta\gamma\delta} = \langle i\mathbf{p}_1\alpha\, j\mathbf{p}_2\beta | k\mathbf{p}_3\gamma\, l\mathbf{p}_4\delta \rangle \delta_{\alpha\delta}\delta_{\beta\gamma} - \langle i\mathbf{p}_1\alpha\, j\mathbf{p}_2\beta | l\mathbf{p}_4\delta\, k\mathbf{p}_3\gamma \rangle \delta_{\alpha\gamma}\delta_{\beta\delta}\,, \tag{4}$$

with

$$\langle i\mathbf{p}_1\alpha\, j\mathbf{p}_2\beta | k\mathbf{p}_3\gamma\, l\mathbf{p}_4\delta \rangle = \int d\mathbf{r}_1 d\mathbf{r}_2 \phi^{*}_{i\mathbf{p}_1\alpha}(\mathbf{r}_1) \phi^{*}_{j\mathbf{p}_2\beta}(\mathbf{r}_2) V(\mathbf{r}_1,\mathbf{r}_2) \phi_{k\mathbf{p}_3\gamma}(\mathbf{r}_2) \phi_{l\mathbf{p}_4\delta}(\mathbf{r}_1)\,, \tag{5}$$

and $a_{i\mathbf{p}\sigma}^{+}$ ($a_{i\mathbf{p}\sigma}$) is the creation (annihilation) operator corresponding to the excitation of electrons (or holes) in *i*-th band with spin σ and momentum **p**. μ is the chemical potential. $\phi_{i\mathbf{p}\alpha}^{*}$ is a single-particle wave-function. Here, we suppose that in Eq.(3), the vertex function consists of the effective interactions between the carriers caused by the linear vibronic coupling in the several bands and the screened coulombic interband interaction of carriers.

When we use the two-band Hamiltonian of Eq.(1) and define the order parameters for the singlet exciton, triplet exciton, and singlet Cooper pair, the mean field Hamiltonian is easily derived [10-17]. Here, we focus three electron scattering processes contributing to the singlet superconducting phase in the Hamiltonian of Eq.(1):

$$g_{i1} = \langle ii|ii\rangle, \quad g_{j1} = \langle jj|jj\rangle, \tag{6}$$

$$g_2 = \langle ii|jj\rangle = \langle jj|ii\rangle, \tag{7}$$

$$g_3 = \langle ij|ij\rangle = \langle ji|ji\rangle, \tag{8}$$

$$g_4 = \langle ij|ji\rangle = \langle ji|ij\rangle. \tag{9}$$

g_{i1} and g_{j1} represent the i -th and j –th intraband two-particle normal scattering processes, respectively. g_2 indicates the intraband two-particle umklapp scattering (Scc Figure 1.)

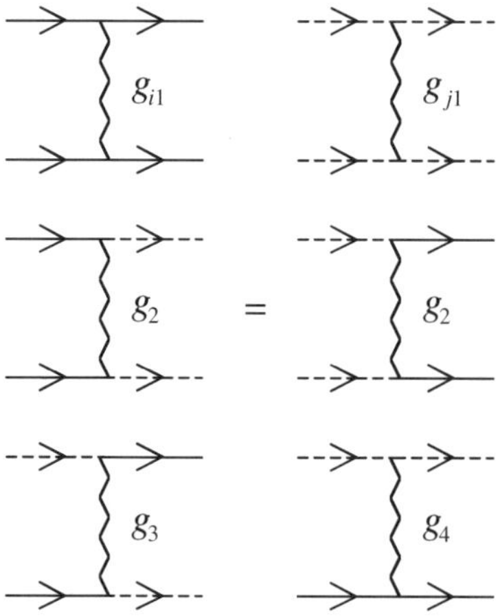

Figure I:4:1. Electron-electron interactions. Solid and dashed lines indicate π- and σ-bands, respectively. g_{i1}, g_{j1}, and g_2 contribute to superconductivity

Note that Γ's are given by

$$\Gamma^{iiii}_{\alpha\beta\gamma\delta} = g_{i1}\left(\delta_{\alpha\delta}\delta_{\beta\gamma} - \delta_{\alpha\gamma}\delta_{\beta\delta}\right),$$
$$\Gamma^{jjjj}_{\alpha\beta\gamma\delta} = g_{j1}\left(\delta_{\alpha\delta}\delta_{\beta\gamma} - \delta_{\alpha\gamma}\delta_{\beta\delta}\right),$$
$$\Gamma^{iijj}_{\alpha\beta\gamma\delta} = \Gamma^{jjii}_{\alpha\beta\gamma\delta} = g_2\left(\delta_{\alpha\delta}\delta_{\beta\gamma} - \delta_{\alpha\gamma}\delta_{\beta\delta}\right),$$
$$\Gamma^{ijij}_{\alpha\beta\gamma\delta} = \Gamma^{jiji}_{\alpha\beta\gamma\delta} = g_3\delta_{\alpha\delta}\delta_{\beta\gamma} - g_4\delta_{\alpha\gamma}\delta_{\beta\delta}, \qquad (10)$$

where an antisymmetrized vertex function Γ is considered to be a constant independent of the momenta. The spectrum is elucidated by the Green function method. Using Green's functions, which characterize the CDW, SDW, and SSC phases, we obtain a self-consistent equation, according to the traditional procedure [10-17]. Then, we can obtain expressions of the transition temperature for some cases.

The spectrum is elucidated by the Green function method. Using Green's functions, which characterize the CDW, SDW, and SSC phases, we obtain a self-consistent equation, according to the traditional procedure [11-15]. Then, we can obtain expressions of the transition temperature for some cases. Electronic phases of a one-dimensional system have been investigated by using similar approximation in the framework of the one-band model [10-17]. In the framework of a mean field approximation with the two-band model, we have already derived expressions of the transition temperature for CDW, SDW, and SSC. In the previous paper [12-15], we have investigated the dependence of T_c on hole or electron concentration for superconductivity of copper oxides by using the two-band model and have obtained a phase diagram of $Bi_2Sr_2Ca_{1-x}Cu_{2x}O_8$ (Bi-2212) by means of the above expressions of transition temperature .

For simplicity, we consider in paper [7] three cases: (1) $g_1 \neq 0$ and others = 0, (2) $g_2 \neq 0$ and others = 0, and (3) g_{i1} and g_{j1}, $g_2 \neq 0$ and others = 0, using of the two-particle Green function techniques. It was shown that for case 3 possibly to arise two superconductivity gap. The superconductivity arising from electron-phonon mechanism $g_1 < 0$ and $g_1 < g_2$) such as MgB$_2$ is in the two-gap region. On the other hand, the superconductivity such as copper oxides ($g_1 > g_2$) is outside the two-gap region. These results predict that we may observe two gap functions for MgB$_2$ and only single gap function for copper oxides.

2.2 Superconductivity in MgB_2

We used case (3) for g_{i1} and g_{j1}, $g_2 \neq 0$ and others = 0 for description superconductivity in MgB_2. We will have reduced Hamiltonian :

$$H = H_0 + H_{\text{int}} \tag{11}$$

where

$$H_0 = \sum_{\mathbf{k},\sigma} \left[[\varepsilon_i - \mu] a^+_{i\mathbf{k}\sigma} a_{i\mathbf{k}\sigma} + [\varepsilon_j - \mu] a^+_{j\mathbf{k}\sigma} a_{j\mathbf{k}\sigma} \right] \tag{12}$$

$$H_{\text{int}} = \sum g_{1i} a^+_{ik} a^+_{i-k} a_{i-k} a_{ik} + \sum i \to j + \sum g_2 a^+_{ik} a^+_{i-k} a_{j-k} a_{jk} \, . \tag{13}$$

Now we define order parameter which are helpful to construct thc mcan field Hamiltonian , defined as

$$\Delta_i = \sum_p \left\langle a^+_{ip\uparrow} a^+_{i-p\downarrow} \right\rangle \tag{14}$$

$$\Delta_j = \sum_p \left\langle a^+_{jp\uparrow} a^+_{j-p\downarrow} \right\rangle \tag{15}$$

The relation between two superconductivity gaps of the system are

$$\Delta_j = \frac{1 - g_{i1} \rho_i f_i}{g_2 \rho_j f_j} \Delta_i \, , \tag{16}$$

where

$$f_i = \int_{\mu}^{\mu - E_c} \frac{d\xi}{\left(\xi^2 + \Delta_i^2\right)^{1/2}} \tanh \frac{\left(\xi^2 + \Delta_i^2\right)^{1/2}}{2T},$$
$$f_j = \int_{\mu - E_c}^{\mu - E_j} \frac{d\xi}{\left(\xi^2 + \Delta_j^2\right)^{1/2}} \tanh \frac{\left(\xi^2 + \Delta_j^2\right)^{1/2}}{2T} \tag{17}$$

with the coupled gap equation:

$$\left(1 - g_{i1}\rho_i f_i\right)\left(1 - g_{j1}\rho_j f_j\right) = g_2^2 f_i f_j. \tag{18}$$

We have tried to estimate the coupling constant of pair electron scattering process between π- and σ-bands of MgB_2 system. We have calculated the parameters by using roughly numerical approximation. We focus one π -band and σ-band of MgB_2 and consider electrons near Fermi surfaces. We find value parameter g_1 = -0.4 eV , using transfer integral between π -band and σ-band .We estimate the coupling parameter g_2 of pair-electron scattering process by the following expression.

$$g_2 = \sum_{k1,k2} V_{k1,k2}^{1,2} \tag{19}$$

$$V_{k1,k2}^{1,2} = \sum_{r,s,t,u} u_{1,r}^*(k1)\, u_{1,s}^*(k1) v_{rs} u_{2,t}(k2) u_{2,u}(k2), \tag{20}$$

where label 1 and 2 mean π- and σ-bands, respectively. $u_{i,r}(ki)$ is LCAO coefficient with *i*-the band and *ki* moment [18,19]. The summation *k1* and *k2* sum over each Fermi surface. However, it is difficult to perform the sum exactly. In this case, we used few point near Fermi surface. The coupling constant of pair-electron scattering between π- and σ-bands is from about g_2=0.025 eV.

From numerical calculations of Eqs.(16)-(18), we can also obtain the temperature dependence of the two gap parameters as shown in Fig.2. We have used the density of states of π- and σ-bands (ρ_i=0.2 eV^{-1}, ρ_j=0.14eV^{-1}), chemical potential μ=-2.0, the top energy of σ-band E_j= -1.0, and fitting parameters (g_{i1}= -0.4 eV, g_{j1}= -0.6 eV, g_2= 0.02 eV).

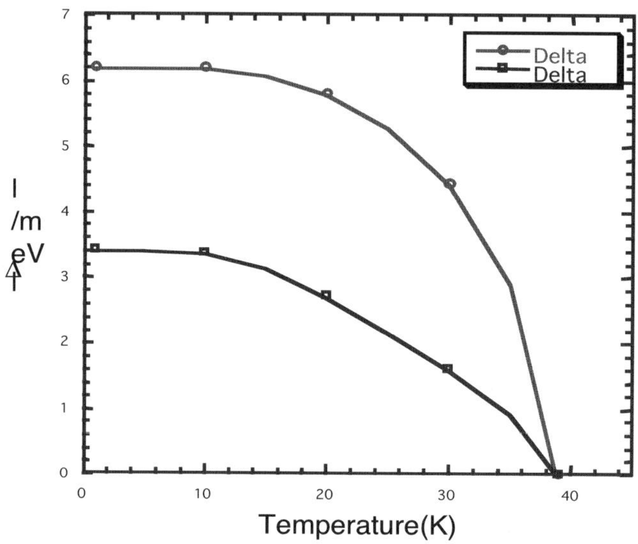

Figure I:4:2. Temperature dependent of two superconductivity gaps

This calculations have qualitative agreement with experiments [20-22].

The expression of transition temperature of superconductivity is derived by a simple approximation :

$$T_{c_+} = 1.13(\zeta - E_j) \cdot \exp\left({-1}/{g_+ \rho} \right) \tag{21}$$

where

$$g_+ = \frac{1}{24}\left(B + \sqrt{B^2 - 4A} \right) \tag{22}$$

and

$$A = g_{1i}g_{1j} - g_2^2 \tag{23}$$
$$B = g_{1i}g_{1j} + g_{1j} \tag{24}$$
$$\zeta = -\mu \tag{25}$$

We can see from expressions for T_{c+} the effect of increase of T_{c+} due to the enhanced interband pairing scattering (g_2)

Figure 3 shows a schematic diagram of mechanism of pairing for two gaps. Scenario is next. Electrons from π and σ zones make up the subsystems. If g_2=0 we have a two independent subsystems with different the transition temperature of superconductivity $T_{c\pi}$ and $T_{c\sigma}$ and two independently superconductivity gaps. In our model we have $g_2 \neq 0$ and two coupled superconductivity gaps with relation (16) and one transition temperature of superconductivity T_{c+} which has agreement with experiments. In our model, we have two channel of superconductivity: conventional channel (intraband g_1) and unconventional channel (interband g_2). Two gaps appear simultaneously in different zones which like BCS gaps. Gap in π zone is bigger to compare with σ zone, because the density of state in π zone is (0.25eV) and in σ zone (0.14eV). Current of Cooper pair gets from π zone in σ zone, because density of Cooper pair in π zone becomes much higher. The tunneling of Cooper pair also stabilizes the order parameter in σ zone.

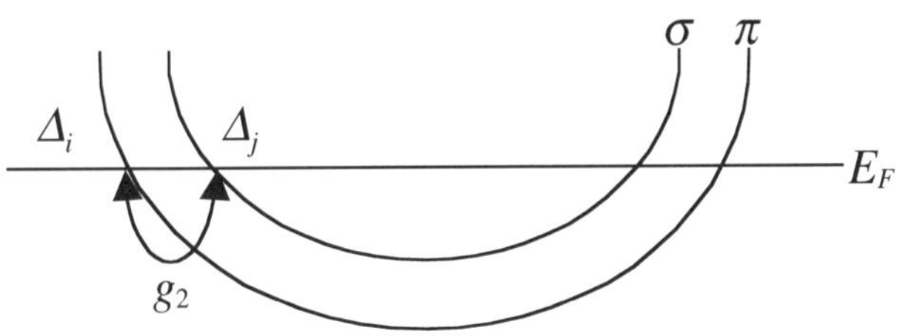

Figure I:4:3. Schematic diagram of mechanism of pairing for two gaps

In this way, we can predict the physical properties of the multi-gaps superconductivity, if we have the superconductors with many zone structure as shown in Fig.3.

3. SUMMARY AND CONCLUSIONS

In conclusion, we have presented our two-band model with intraband two-particle scattering and interband pairing scattering processes to describing two-gap superconductivity in MgB_2.

We defined the parameters of our model and made numerical calculations of temperature dependent of two gaps. It is a qualitative agreement with experiments. We proposed a two channel scenario of superconductivity: first a conventional channel (intraband g_1) whichis is connected with BCS mechanism in different zone and a unconventional channel (interband g_2) which describes the tunneling of a Cooper pair between two bands. The tunneling of Cooper pair also stabilizes the order parameters of superconductivity [9-12] and increases the critical temperature of superconductivity.

ACKNOWLEDGMENT

S. Kruchinin is grateful for financial support by the grant INTAS – 654. H. Nagao is grateful for a financial support of the Ministry of Education, Science and Culture of Japan (Research No.15550010, 15035205).

REFERENCES

1. J. Nagamatsu, N. Nakamura, T. Muranaka, Y. Zentani, and J. Akimitsu, *Nature*, 410, 63 (2001).
2. A. Bianconi, and M. Filippi, this volume Chapter 2; A. Bianconi, et al. *J. Phys.: Condens. Matter* 13, 7383 (2001); J. P.C. Canfield, S.L. Bud'ko and D.K. Finnemore, *Physica C,* 385, 1 (2003).
3. H. Suhl, B.T. Matthias, R.Walker, *Phys.Rev.Lett.,* 3, 552 (1959).
4. V.A. Moskalenko, *Fiz.Metalloved.*8, 25, (1959)
5. J. Kondo, *Prog . Theor. Phys.* 29, 1 (1963)
6. G. Binnig , A. Baratoff, Hoening, J.G. Bednorz , *Phys.Rev.Lett.* 45, 1352, (1980)
7. J. M. An and W. E. Picket, *Phys. Rev. Lett.* 86, 4366 (2001).
8. J. Kortus, I.I. Mazin, K.D. Belashenko, V.P.Antropov and I.L.Boyer, *Phys.Rev.Lett.* 86, 4656 (2001).
9. J. G. Bednorz and K. A. Müller, *Z. Phys. B* , **64**, 189 (1986).

10. H. Nagao, S.P. Kruchinin, A.M. Yaremko, and K.Yamaguchi, *Int. J. Mod. Phys. B,* **16,** 23, 3419 (2002) .
11. H. Nagao, M. Nishino, Y. Shigeta, Y. Yoshioka and K. Yamaguchi, *J. Chem. Phys.,* 113, 11237 (2000).
12. H. Nagao, H. Kawabe, S.P. Kruchinin, D.Manske , K. Yamaguchi, *Mod.Phys.Lett. B*, **17**, 10 , 423, (2003) .
13. H. Nagao, A.M. Yaremko, S.P. Kruchinin, and K.Yamaguchi, in J. Annett and S. Kruchinin (eds.), *New Trends in Superconductivity,* Kluwer Academic, Dordrecht, 155 (2002)
14. H. Nagao, S.P. Kruchinin , and K.Yamaguchi, in J.K. Srivastava and M. Rao (eds.), *Models and Methods of High-Tc superconductivity: Some Frontal Aspects,* Nova Science Publishers, Inc. Academic, 205 (2003)
15. H. Nagao, Y. Kitagawa, T. Kawakami, T. Yoshimoto, H. Saito, and K. Yamaguchi, *Int. J. Quantum Chem.*, 85, 608 (2001).
16. A.M. Yaremko,E.V. Mozdor, and S.P. Kruchinin, *Int. J. Mod. Phys.* B **10**, 2665 (1996).
17. M. Yaremko, E.V. Mozdor, H. Nagao, and S.P. Kruchinin, in J. Annett and S. Kruchinin, (eds), *New Trends in Superconductivity ,* Kluwer Academic , Dordrecht , 329-339.
18. M. Yamamoto, K. Nishikawa, S. Aono, Bull. Chem. Soc. Jpn, 58, 3176, (1985)
19. M. Kimura, H. Kawabe, A. Nakajima, K. Nishikawa, S. Aono, Bull. Chem. Soc. Jpn., 61, 4239 (1988)
20. T. Örd , N. Kristoffel, and K. Rägo, in S. Kruchinin (eds.), *Modern Problems of superconductivity, Mod. Phys. Let.* B , 17, 667 (2003).
21. N. Kristoffel, T. Örd, K. Rägo, *Europhys. Lett.*, 61,109 (2003). P. Szabo et al., *Phys. Rev. Lett.* 87, 137005 (2001).

II

ANOMALOUS ELECTRON-PHONON INTERACTION

II.1

ELECTRON-LATTICE COUPLING IN THE CUPRATES

T. Egami
University of Tennessee, Knoxville, TN 37996, USA and Oak Ridge National Laboratory, Oak Ridge, TN 37831, USA

Abstract: This article summarizes the principal points of discussions at the Erice workshop on the relevance of electron-lattice coupling to high temperature superconductivity. While the majority in the field still believes that the phonons and lattice are irrelevant to the superconductivity in the cuprates, such a view is strongly challenged by recent experimental results, particularly in conjunction with the increasing evidence of intrinsic electronic inhomogeneity. It is likely that phonons are a vital component of the phenomenon through the unconventional electron-phonon coupling.

1. INTRODUCTION

It is now a widely accepted paradigm that the remarkable properties of transition metal oxides originate from the closely coupled interplay of spin, charge and lattice degrees of freedom. For instance the colossal magnetoresistance (CMR) in manganites is a consequence of an insulator-to-metal transition (IMT) induced by a magnetic field. In this system the IMT is concomitant with a ferromagnetic transition, and is determined by a precarious balance between the force to localize the charge (due to the electron-lattice interaction and spin correlation) and the force to delocalize the charge (due to the electron kinetic energy and elastic energy). When the

A. Bianconi (ed.), Symmetry and Heterogeneity in High Temperature Superconductors, 79–86.

localization force is dominant charges are localized as spin-lattice polarons, and the system shows an insulating behavior. Thus the electron-lattice interaction is critically important, as first pointed out by Millis *et al.* [1] and was proven by further research.

In stark contrast, in the research field of the high temperature superconductivity (HTSC) the role of the lattice has been all but completely neglected by the majority. The conventional view is that it is a purely electronic phenomenon involving spin excitations, and is described, for instance, by the *t-J* model [2]. There are many reasons why the lattice has been dropped from consideration almost from the beginning, such as the near absence of the isotope effect on the critical temperature, T_C, and the linear resistivity. However, the arguments against the lattice involvement are less than perfect [3].

In the meantime evidence of significant lattice involvement has steadily been accumulating. The discussions at this workshop covered recent observations by ARPES, isotope effect and neutron scattering. At the same time recent theories suggest that the electron-phonon (*e-p*) coupling in the strongly correlated electron systems is unconventional, involving spins. Conventional arguments against the phonon mechanism are based on the *e-p* coupling in the Fermi liquids, and may not apply to the cuprates. For a long time since the discovery of the HTSC in 1986 the researchers who take phonons seriously have been a definite minority. However, while other theories have not been able to survive increased level of experimental scrutiny the relevance of phonons appears to be taking an opposite, rising trail. The majority may have made a premature decision on the phonon before important data were obtained.

2. EXPERIMENTAL OBSERVATION

A large amount of experimental results show, rather convincingly, that certain features of the local structure, not the average structure, respond to the superconductive transition or to the opening of a pseudo-gap [4]. The majority dismissed them as mere consequences of strong superconductivity, but more recent data provide more direct evidence of significant lattice involvement.

The recent results of angle-resolved photo-electron spectroscopy (ARPES) by A. Lanzara *et al.* show quite convincingly that the phonons, most likely the zone-boundary oxygen LO phonons, interact strongly with electrons [5,6]. The position of the kink in the electron dispersion along the nodal direction (π, π) is largely independent on composition and the magnitudes of the gap and T_C, but changes with isotope substitution from ^{16}O to ^{18}O. Along the anti-nodal direction (π, 0), they found that the electron dispersion shifts significantly with oxygen isotope substitution. Oxygen isotope substitution changes the volume very slightly, since ^{16}O has a larger zero-point phonon amplitude than ^{18}O. However, this volume effect does not appear to be large enough to explain the observed shifts. It is most likely that this shift is another evidence of strong *e-p* coupling. While the coupling along the nodal direction does not contribute to superconductivity the coupling along the anti-nodal direction does. This observation has increased the possibility that the phonons contribute to the HTSC phenomenon.

The superconductive penetration depth was found to depend strongly on the oxygen isotopic mass by the group of H. Keller, and the result was interpreted to represent the isotope effect on the electronic effective mass [7]. This observation ties in well with the ARPES observation mentioned above, and provides another piece of evidence for the strong *e-p* coupling and its relevance to the HTSC. While it may be argued that the observation provides a definite proof that the phonons significantly modify the electron dispersion, and may not directly prove that they contribute to the pairing, it is difficult to imagine that such a strong interaction does not have an important role in pairing.

The phonon branch that is suspected to interact strongly with the charge by the ARPES measurement, namely the zone-boundary Cu-O bond-stretching LO phonon branch (Fig. 1), was found by inelastic neutron scattering to show unusual temperature dependence [8].

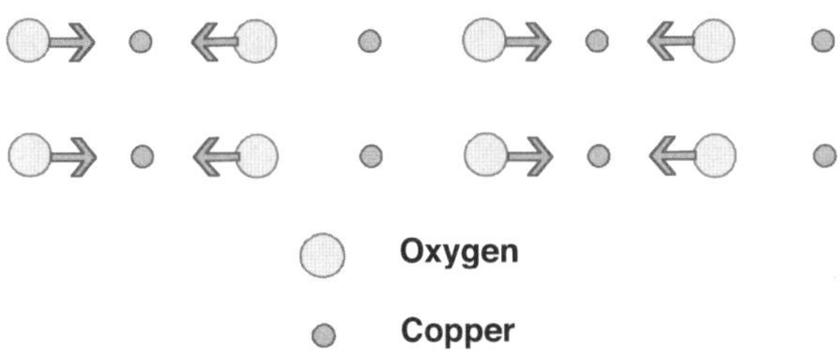

Figure II:1:1 Schematics of the Cu-O bond-stretching zone-boundary LO phonon (half-breathing) mode

The intensity of inelastic neutron scattering from this branch changes with temperature just as the superconductive order parameter. At the same time other phonon branches with polarization along the c-axis, such as the apical oxygen mode and the buckling mode, do not show appreciable temperature dependence. This suggests that the zone-boundary oxygen LO phonons have a special significance to the lattice effect in the cuprates.

3. THEORETICAL PERSPECTIVE

The reason why the Cu-O bond-stretching phonons are important is not difficult to understand. The undoped cuprates are charge-transfer insulators [9]. The oxygen *p*-level and the Cu *d*-level form hybridized anti-bonding orbital states, and the gap lies between the filled *p*-like band and the empty upper Hubbard Cu-*d*-like band. When the Cu-O bondlength is modified the extent of the *p*-*d* hybridization is modified. This situation is somewhat similar to that of ferroelectric oxides, such as $BaTiO_3$. In $BaTiO_3$ the insulating gap is between the filled *p*-like state and the empty *d*-like state, and shortening the Ti-O distance increases the *p*-*d* hybridization, and thus results in the transfer of negative charge from the *p*-state of O to the *d*-state of Ti. A very rigorous and elegant theory was recently worked out by Vanderbilt and Resta [10,11], which relates this charge transfer to the electronic polarization in the compound that is as large as the ionic polarization, doubling the Born effective charge in many cases.

In the doped cuprate, as it turned out, shortening the Cu-O distance results in the transfer of ***positive*** charge, since the doped holes reside mostly on the oxygen *p*-like band. Thus the electronic polarization is antiparallel to the ionic polarization, greatly reducing the Born effective charge, or even changing its sign. As shown in Fig. 2 for the case of a one-dimensional Hubbard model, the effective charge strongly depends on the phonon wavevector, q, and becomes maximum in the middle of the Brillouin zone for the doping level, $x = 1/4$ [12]. This strong charge transfer results in the phonon softening [13,14], and strong *e-p* coupling.

It was found that because of the strong electron correlation the phonon-induced charge transfer depends on the spin. Consequently the phonon modifies the spin correlation and spin dynamics [12]. In the strongly correlated electron systems phonons interact not only with charge, but also with spins.

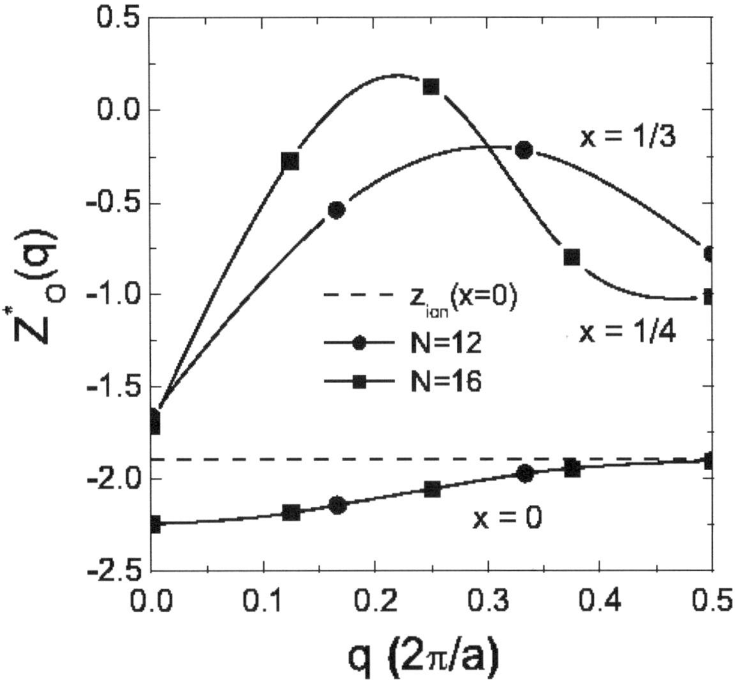

Figure II:1:2. The Born effective charge of oxygen calculated for the one-dimensional two-band Hubbard model with the Cu-O bond-stretching LO mode, calculated for N = 12 ring with doping level x = 0, 1/3 and for N = 16 ring with x = 0, 1/4. The dashed line correspond to the static (ionic) charge [12]

This is why we call the *e-p* interaction in the cuprates ***unconventional***. In addition, because the phonon induces charge transfer from O to Cu over the charge transfer gap of 2 eV, the effect of the phonon extends to high energy ranges [12]. The recent observation that the depression in the optical spectrum due to the gap opening extends to 2 eV was used as the basis for the argument for the electronic origin of the HTSC phenomenon [15-17]. However, it can be explained by the phonon-induced charge transfer [12].

Because of the topology of the CuO_2 plane this *e-p* coupling is anisotropic, strong along the (π, 0) direction and weak along the (π, π) direction. The (π, π) mode is a full breathing mode that localizes charge on Cu. The kinetic energy penalty of localization makes this mode less strongly coupled to charge. The (π, 0) mode, on the other hand, is "half-breathing", and allows charge to flow in the perpendicular direction, thus lowering the kinetic energy penalty. Because of this anisotropy the phonons are even capable of supporting the *d*-wave superconductivity [18], while the *s*-wave solution is usually preferred for the phonon mechanism.

4. RELATION TO SPATIAL INOMOGENEITY AND RELEVANCE TO HTSC

Even though the ARPES results show strong *e-p* coupling, the values of the *e-p* coupling parameter, λ, estimated from the ARPES data along the nodal direction, 1-2, is not large enough to bring about the HTSC. At the same time the spatial inhomogeneity of HTSC has long been suspected [19,20], and is now recognized as one of the common features of the HTSC cuprates as discussed in the other section in this volume [21]. Is it just an accident that the HTSC cuprates are inhomogeneous, or could it be a part of the mechanism? To answer this question we should examine the composition dependence of the pseudo-gap temperature, T_{PG}. This is a complex and confusing problem, but the T_{PG} line that lies above and covers the entire T_C dome seems to coincide with the start of local structural distortion [22]. In particular the fall of T_C on the overdoped side coincides with T_{PG}. This suggests that the spatial inhomogeneity of spin and charge and local lattice distortion associated with such inhomogeneity are the ***required*** conditions for HTSC. Without the phase separation the overdoped cuprate is just a regular Fermi liquid with a zero or vanishingly small value of T_C.

This is a very unique situation for superconductivity, since in the previous experience inhomogeneity is almost always harmful to superconductivity. Why is the superconductivity in the cuprates so different? While further research is clearly required to answer this puzzle, one possibility is that the spatial confinement produces the vibronic resonant state of phonon and charge that enhances HTSC [15,23]. The benefit of spatial confinement on HTSC has been strongly advocated for some time by Phillips with the idea of filamental superconductivity [24] and more recently by Bianconi [25] as the shape resonance effect. In both cases the effect arises due to the enhancement of the local density of states (DOS). An additional, and possibly more central, effect of confinement is to reduce the group velocity of electrons and bring it comparable to the phonon velocity, thus making the non-adiabatic vibronic resonance possible [15,23]. While details need further scrutiny, the enhancement of the unconventional *e-p* coupling by spatial inhomogeneity appears to be the crucial element of the HTSC mechanism.

5. CONCLUDING REMARKS

Phonons have been discredited and ignored from the very early days of the HTSC research, particularly by theoreticians. However, such hasty dismissal always had a smell of avoidance for convenience without careful scientific examination, a guilty verdict without a due process. This is understandable since phonons are an old story, even called the "conventional mechanism". For theoreticians who thrive at the novelty of ideas working with phonons do not bring needed excitement and the possibility of glory. However, the experimental results have started to penetrate the cold wall of rejection and to chip away misgivings and misconceptions about the importance of the lattice effects. In addition it appears that the phonons in the cuprates are not "conventional" after all. They couple to spins, and may resonate with charge in a non-adiabatic vibronic manner due to magnetic and structural confinement. In this case we are not dealing with the old BCS mechanism any more, but with a drastically new and unexplored HTSC mechanism, worth a serious study. While the jury is still out on the relevance of the lattice to the HTSC mechanism, we may have succeeded in obtaining the court order for the stay of execution of the hastily issued, probably wrong, earlier verdict.

ACKNOWLEDGMENT

The author is grateful to the panel members for powerful and convincing presentations and the audience for the insightful discussion. This work was supported by the National Science Foundation through DMR01-02565.

REFERENCES

1. A. J. Mills, P. B. Littlewood, and B. I. Shraiman, Phys. Rev. Lett. **74**, 5144 (1995).

2. *e.g.*, P. W. Anderson, *The Theory of Superconductivity in the High-T_C Cuprates* (Princeton University Press, Princeton, 1997).
3. P. B. Allen, Nature (London), **412**, 494 (2001).
4. T. Egami and S. J. L. Billinge, in *Physical Properties of High Temperature Superconductors V*, ed. D. Ginsberg (Singapore, World Scientific, 1996) p. 265.
5. A. Lanzara *et al.*, Nature (London), **412**, 510 (2001).
6. G.-H. Gweon, T.Sagawa, S.Y. Zhou, J. Graf, H. Takagi, D.H. Lee and A. Lanzara, Nature (London) **430**, 187 (2004)
7. R. Khasanov, *et al.*, *Phys. Rev. B* **68**, 220506 (2003); *J. Phys. Cond. Mat.* **15,** L17 (2003).
8. J.-H. Chung, T. Egami, R. J. McQueeney, M. Yethiraj, M. Arai, T. Yokoo, Y. Petrov, H. A. Mook, Y. Endoh, S. Tajima, C. Frost, and F. Dogan, *Phys. Rev. B*, **67**, 014517 (2003).
9. J. Zaanen, G. A. Sawatzky and J. W. Allen, *Phys. Rev. Lett.* **55**, 418 (1985).
10. R. D. King-Smith and D. Vanderbilt, *Phys. Rev. B* **47**, 1651 (1993).
11. R. Resta, *Rev. Mod. Phys.* **66**, 899 (1994).
12. P. Piekarz and T. Egami, *Physica* C**408-410**, 334(2004)
13. S. Ishihara, T. Egami and M. Tachiki, *Phys. Rev. B* **55**, 3163 (1997).
14. Y. Petrov and T. Egami, *Phys. Rev. B* **58,** 9485 (1998).
15. D. N. Basov, E. J. Singley and S. V. Dordevic, *Phys. Rev. B* **65**, 054516 (2002).
16. H. J. A. Molegraaf, C. Presura, D. van der Marel, P. H. Kes and M. Li, *Science* **295**, 2239 (2002).
17. A. F. Santander-Syro, R.P.S.M. Lobo, N. Bontemps, Z. Konstantinovic Z, Z. Z. Li, H. Raffy, *Eur. Phys. Lett.* **62**, 568 (2003).
18. M. Tachiki, M. Machida and T. Egami, *Phys. Rev. B*, **67**, 174506 (2003).
19. L. P. Gorkov and A. V. Sokol, *JETP Lett.* **46**, 420 (1987).
20. J. C. Phillips, *Phys. Rev. Lett.,* **59**, 1856 (1987).
21. *e.g.*, K. M. Lang, *et al.*, *Nature* **415**, 412 (2002).
22. *e.g.*, N. L. Saini, *et al.*, *Eur. Phys. J. B*, **36,** 75 (2003).
23. J. B. Goodenough and J. Zhou, *Phys. Rev. B*, **42**, 4276 (1990).
24. J. C. Phillips, *Phys. Rev. B,* **41,** 8968 (1990).
25. *e.g.*, A. Bianconi, *Int. J. Mod. Phys. B*, **14**, 3289 (2000)

II.2

SYMMETRY BREAKING, NON-ADIABATIC ELECTRON-PHONON COUPLING AND NUCLEAR KINETIC EFFECT ON SUPERCONDUCTIVITY OF MgB_2

P. Baňacký

Comenius University, Faculty of Natural Science, Institute of Chemistry, Chemical Physics division. 84215 Bratislava, and S-Tech a.s., Dubravska cesta 9, 84219 Bratislava, Slovakia

Abstract: Theory of non-adiabatic electron-vibration interactions has been applied to the study of MgB_2 superconductivity. It has been shown that at the non-adiabatic conditions when the Born-Oppenheimer approximation is not valid, and electronic motion is dependent not only on the nuclear coordinates but also on the nuclear momenta, the fermionic ground state energy of some systems can be stabilised by electron-phonon interactions at broken translation symmetry and an energy gap in one-particle metalic spectrum can be opened. More over, the new arising state is geometricaly degenerate – i.e., there is an infinite number of different nuclear configurations with the same fermionic ground state enegy. The superconducting state transition can be characterised as a Non-Adiabatic Sudden Increase of the Cooperative Kinetic Effect at Lattice Energy Stabilization (NASICKELES). Comparing to the experiments, the model study of MgB_2 yields very good results. Calculated T_c is 39.5 K, and density of states exhibits two-gap character in full agreement with the tunneling spectra. The peaks are at +/-4 meV that is connected to π band and at +/-7.6 meV for σ band

A. Bianconi (ed.), Symmetry and Heterogeneity in High Temperature Superconductors, 87–101.

1. INTRODUCTION

Comparable values of the Fermi energy and energy of the relevant vibration/phonon mode(s), i.e. $E_F \approx \leq \omega$, indicate that electronic and nuclear motions are mutualy dependent and cannot be studied as statistically independent fields. Consequently, separation of the electronic and nuclear motions by means of the Born – Oppenheimer (BO) approximation (or application of Migdal theorem to a theory of superconducting state transition) with the correction to the electronic energy over the diagonal part of the nuclear energy term (diagonal BO correction – DBOC) and corrected nuclear potential energy, i.e. adiabatic approximation, is for such systems incorrect.

It means that standard field theory Hamiltonian of the form,

$$\mathrm{H} = \mathrm{H}_{ee} + \mathrm{H}_{ph} + \mathrm{H}_{int} \approx \sum_k \varepsilon_k a_k^+ a_k + \sum_v \hbar w_v (b_v^+ b_v + \tfrac{1}{2}) + H_{int}(a^+ a, b^+ b)$$

is an inadequate starting point for correct description of such systems.

1.1 Non-adiabatic treatment of electronic structure calculation

Starting Hamiltonian is written in the general form of nonrelativistic – "chemical" Hamiltonian of interacting set of atoms ,

$$H = T_N(P) + E_{NN}(Q) + \sum_{PQ} h_{PQ}(Q) a_P^+ a_Q + \frac{1}{2} \sum_{PQRS} v_{PQRS}^0 a_P^+ a_Q^+ a_S a_R \qquad (1)$$

The Hamiltonian is an explicate function of the operators of nuclear coordinates, $\bar{Q}_V$, and nuclear momenta, $\bar{P}_V : \bar{Q}_V = (\bar{b}_V^+ + \bar{b}_V)$, $\bar{P}_V = (\bar{b}_V - \bar{b}_V^+)$

The terms, $E_{NN}(\bar{Q})$ and $h_{PQ}(\bar{Q})$ are introduced over the Taylor's expansion at the equilibrium geometry of fixed nuclear configuration R_0 (crude level – clamped nuclei),

$$E_{NN}(\bar{Q}) = E_{NN}^O(eq) + \sum_{i=1}^{\infty} E_{NN}^{(i)}(\bar{Q}) \quad , \quad h_{PQ}(\bar{Q}) = h_{PQ}^0 + \sum_{i=1}^{\infty} u_{PQ}^{(i)}(\bar{Q})$$

$$u_{PQ}(\bar{Q}) = \left\langle P / \sum_j \frac{-Z_J e^2}{/r - R_j /} / Q \right\rangle * f(\bar{Q})$$

The term $E^0_{NN}(eq)$ is the potential energy of the clumped nuclei configuration at the equilibrium geometry R_0 (crude level), h^0_{PQ} is one – electron term ("core" Hamiltonian) for equilibrium nuclear configuration R_0, and $u_{PQ}(\bar{Q})$ represents matrix element of electron – vibration (phonon) coupling.

The problem introduced by this Hamiltonian is studied by means of the unitary – canonical quaziparticle transformation technique in two steps [1],

1st. step: adiabatic, Q-dependent transformation,

$$\bar{\bar{H}}_a = e^{-S_1(Q)} H e^{S_1(Q)} \tag{2}$$

2nd. step: non-adiabatic, P-dependent transformation,

$$H_{na} = e^{-S_2(P)} \bar{\bar{H}}_a e^{S_2(P)}, \tag{3}$$

with subsequent iterative solution of the resulting coupled set of SCF-HF, GCPHF (generalized coupled perturbed Hartree-Fock, which mix electronic and nuclear motion) and nuclear secular equations. The procedure is connected to geometry optimization of the system with respect to the total energy of system.

The solution yields [1], beside the final set of normal modes frequencies, the final nonadiabatic form of the fermionic Hamiltonian, that is:

$$H_F = H_A(crude) + \Delta H_F \tag{4}$$

In this equation, H_A(crude) is the clamped nuclei (crude – BO level) Hamiltonian,

$$H_A(crude) = H_{A^0(0)} + H_{A'(0)} + H_{A''(0)} \tag{5}$$

where $H_{A^0(0)}, H_{A^\S(0)}, H_{A^{\S\S}(0)}$ are: zero-particle term (clumped nuclei fermionic ground state energy), one-particle Hamiltonian (one-particle spectrum - orbital energies) and two-particle Hamiltonian (correlation energy).
The non-adiabatic corrections to the electronic energy are represented by the term ΔH_F ,

$$\Delta H_F = \Delta H_F^0 + \Delta H_F' + \Delta H_F'' \tag{6}$$

with the corresponding corrections to zero, one and two – particle terms.
Under the assumption of the dominance of electron–nuclear interaction (one-particle terms) at the coupling of the electron-nuclear motion, an approximate analytic solution of the GCPHF equations [1] for adiabatic Q-dependent (C_{PQ}^r) and non-adiabatic P-dependent ($\hat{C}_{PQ}^r$) unitary transformation coefficients is:

$$C_{PQ}^r = U_{PQ}^r \frac{(\varepsilon_P^0 - \varepsilon_Q^0)}{(\hbar w_r)^2 - (\varepsilon_P^0 - \varepsilon_Q^0)^2} ; \hat{C}_{PQ}^r = U_{PQ}^r \frac{\hbar w_r}{(\hbar w_r)^2 - (\varepsilon_P^0 - \varepsilon_Q^0)^2}$$

The basic equations of the nonadiabatic corrections to electronic energy are:
1. Ground – state energy correction (real - space orbital representation),

$$\Delta H_F^0 = \Delta E_F^0 = \sum_{AIr} \hbar w_r (|C_{AI}^r|^2 - |\hat{C}_{AI}^r|^2) = \sum_{AI} \Omega_{AI} \tag{7}$$

{A} – unoccupied molecular orbitals, {I} – occupied molecular orbitals, {r} – normal modes, with Ω_{PQ} being a symmetric matrix,

$$\Omega_{PQ} = \sum_r \hbar w_r (|C_{PQ}^r|^2 - |\hat{C}_{PQ}^r|^2) = \sum_r \hbar w_r \frac{|U_{PQ}^r|^2}{(\varepsilon_P^0 - \varepsilon_Q^0)^2 - (\hbar w_r)^2} = \Omega_{QP}$$

For k – space representation this equation reads,

$$\Delta H_F^0 = \Delta E_F^0 = 2 \sum_{k,k'} |U^{k-k'}|^2 f_k (1 - f_{k'}) \frac{\hbar w_{k'-k}}{(\varepsilon_{k'}^0 - \varepsilon_k^0)^2 - (\hbar w_{k'-k})^2} \tag{8}$$

2. One – particle energy correction:
a) Correction to the orbital energies (part of the one-particle correction that influence shift of the orbital energies with respect to E_F - i.e. boson operators independent part). The real space orbital representation is,

$$\Delta H_F^{'} = \sum_P \Delta\varepsilon_P N\left[a_P^{+} a_P\right] = \sum_P (\sum_A \Omega_{PA} - \sum_I \Omega_{PI}).N\left[a_P^{+} a_P\right] \tag{9}$$

b) The total one-particle corrections due to nonadiabatic electron- vibration interactions (includes the above boson-independent contribution and boson dependent part – expression below is for boson vacuum):
k – space representation ("dressed polarons"),

$$\Delta H_F^{'} = \sum_{k,q,\sigma} \left|U^q\right|^2 \frac{1}{(\varepsilon_k^0 - \varepsilon_{k-q}^0) - \hbar w_r} N\left[a_{k,\sigma}^{+} a_{k,\sigma}\right] -$$

$$-2.\sum_{k,q,\sigma} \left|U^q\right|^2 f_{k-q} \frac{\hbar w_q}{(\varepsilon_k^0 - \varepsilon_{k-q}^0)^2 - (\hbar w_q)^2} N\left[a_{k,\sigma}^{+} a_{k,\sigma}\right] \tag{10}$$

In this expression, the first term is basicaly an energy of single polaron ("Lee – Low – Pines" polaron) and the second term is an energy correction to a polaron energy due to SCF field of all other polarons.

3. Two - particle energy correction: correction to electron – electron correlation energy due to the phonon field. This non-adiabatic term represents full attractive contribution, and can be compared to the reduced form of Fröhlich effective Hamiltonian which maximizes attractive contribution of electron – electron interaction and that can be either attractive or repulsive (interaction term of the BCS theory). For superconducting state transition at the non-adiabatic conditions, the two-particle correction is unimportant – see [2].

2. NON-ADIABATIC EFFECTS IN MgB_2: RESULTS

The non-adiabatic theory indicates that electron - vibration (phonon) interaction, at stabilization (minimization) of the fermionic ground state energy, can induce nuclear displacements out-of the original equilibrium configuration R_0, i.e. to break the original translation symmetry [3]. At these circumstances, it is necessary to investigate if for MgB_2 there exists a distorted structure that is more stable, due to non-adiabatic EP interactions, than the original equilibrium structure, and look for possibility of some other then BCS mechanism at the transition to superconducting state.

2.1 Band structures of MgB_2

Presented band structure calculations have been done by computer code SOLID 2000 [4,5]. The lattice parameters of MgB_2 (hexagonal structure, space group P6mmm), with the fraction coordinates of the unit cell atoms: Mg = (0,0,0); B1 = (1/3,2/3,1/2); B2 = (2/3,1/3,1/2), have been optimized in a good agreement with the experiment.

a/ Equilibrium – undistorted geometry

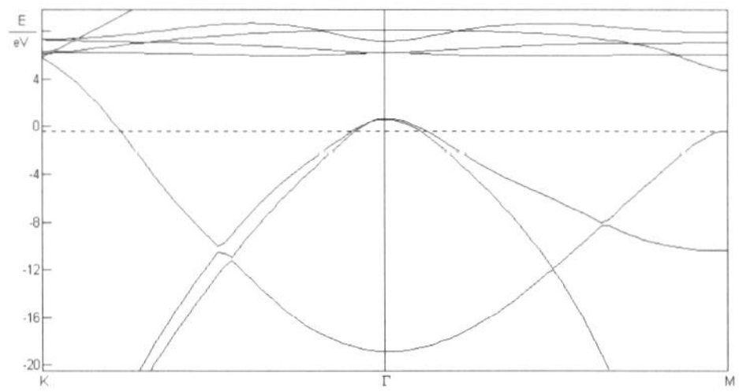

Figure II:2:1. Band structure of MgB_2 calculated at the equilibrium – undistorted geometry - details for the path K-Γ-M. The Fermi level - E_F is indicated by the dashed line

b/ Distorted geometries

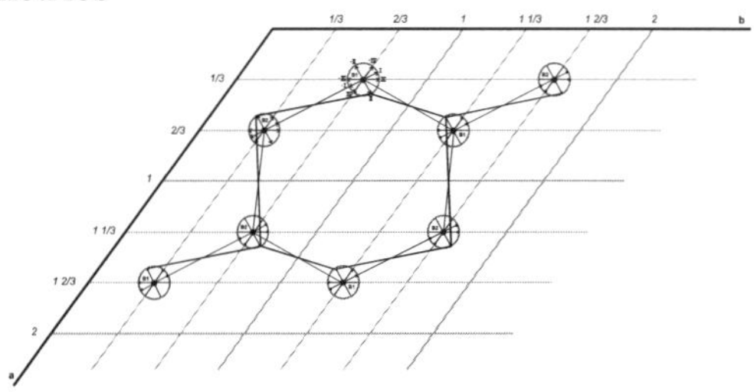

Figure II:2:2. undistorted high symmetry structure of B atoms in **a-b** plane and symmetry breaking by the in-plane, out-of phase displacements $|\Delta f|$/atom. The motion of B1-B2 atoms in out-of phase positions along the perimeters of circles with radius Δf centred at the undistorted B atoms positions generates an infinite number of distorted structures. The figure indicates beside the undistorted structure also one of the distorted structures that correspond to the E_{2g}(b) phonon mode distortion (bold line)

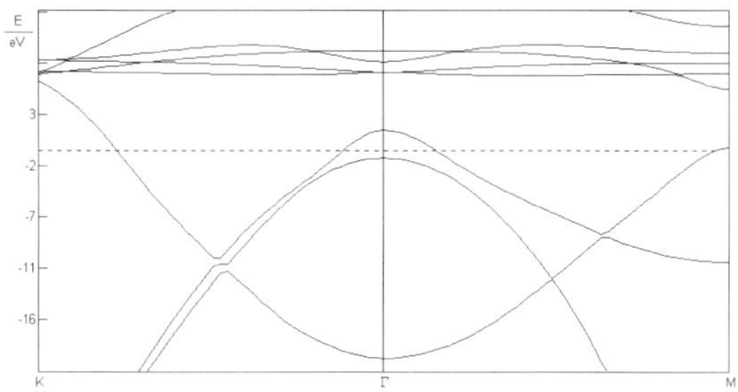

Figure II:2:3. Band structure of MgB_2 calculated for the distorted geometry. The displacement of B atoms is $|\Delta f| = 0.005$/atom (fraction unit) that correspond to 0.032 A° of B1-B2 bond length elongation for stretching vibration (E_{2g}(a) phonon mode amplitude). At this displacement, the lower splitoff σ band has just sank below E_F. The Fermi level - E_F is indicated by the dashed line.The band structure calculations indicate that dominant is σ_1-σ_2 and σ_1-π bands coupling over the E_{2g} phonon mode

2.2 Deformation energy

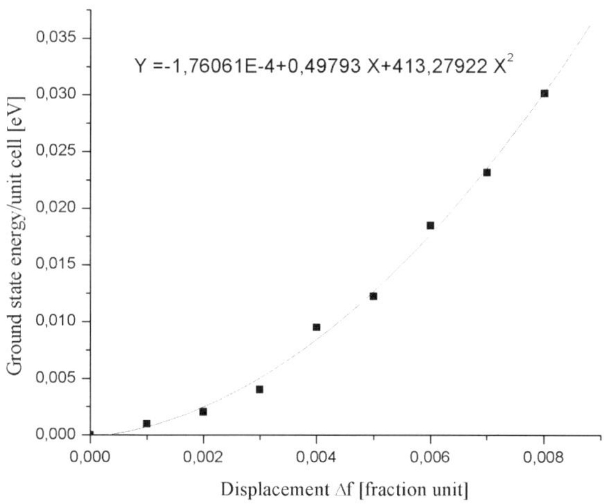

Figure II:2:4. The MgB_2 electronic ground state energy/unit cell as the function of the displacement Δf. The electronic energy of the undistorted structure is the reference value - 0 eV. The dependence is harmonic over the studied range of the displacements, and is identical for all studied types of the distortions. At the displacement $|\Delta f| = 0.005$/atom, the lower splitoff σ band has just sank below E_F, and the ground state energy has been destabilized by 12 meV

2.3 Non-adiabatic correction to zero – particle term of the fermionic Hamiltonian. Correction to the fermionic ground state energy

The study of the E_{2g} phonon mode dispersion [6] indicates different frequencies at Γ and K points, $\hbar\omega_\Gamma \approx 0.066$ eV , and $\hbar\omega_K \approx 0.090$ eV. In this case, for the fermionic ground state energy correction holds,

$$\Delta H_F^0 = \Delta E_F^0 = VP \int_{z_{\min}}^{z_{\max}} \left| U_{z,\Gamma}^{(E_{2g})} \right|^2_{\sigma-\sigma} n_{z,\sigma} \frac{\hbar\omega_\Gamma}{z^2 - (\hbar\omega_\Gamma)^2} dz + VP \int_{z_{\min}}^{z_{\max}} \left| U_{z,K}^{(E_{2g})} \right|^2_{\sigma-\pi} n_{z,\pi} \frac{\hbar\omega_K}{z^2 - (\hbar\omega_K)^2} dz \qquad (11)$$

with z being, $z = (\varepsilon_A^0 - \varepsilon_I^0)$, i.e., the difference of the orbital energies of interacting pairs of unoccupied states $\{\varepsilon_A^0\}$ and occupied states $\{\varepsilon_I^0\}$ of the particular clumped nuclear structure. Integration in above equation runs from $z_{\min} = \hbar\omega_r/q$ up to $z_{\max} >> \hbar\omega_r (= \infty)$. The quantity "**q**" ($q \geq 1$) in $z_{\min}$ "regulates" the distance of the top of the lower splitoff σ band from E_F. The quantity n_z represents „density" for energy differences z of the interacting pairs of states.The matrix element of EP coupling for phonon mode r, $U_z^{(r)}$, is also z-dependent.

$\bar{U}_{\sigma-\sigma} = \left\langle U^{(E_{2g})} \right\rangle_{\sigma=\sigma} \approx 0.7$ eV , for σ-σ bands interaction, and

$\bar{U}_{\sigma-\pi} = \left\langle U^{(E_{2g})} \right\rangle_{\pi=\sigma} \approx 0.25$ eV , for π-σ bands interaction.

The ground state energy loss (destabilization) due to symmetry distortion, for the displacement $|\Delta f| = 0.005$, when the top of the lower splitoff σ band has approached, crossed E_F and sank just below it, has been calculated to be

+12 meV/unit cell (see Fig. 4). For this value of the displacement, the ground state energy correction due to non-adiabatic EP coupling is about –50 meV (Fig.6.,q = 2). Due to k : (-k) symmetry, this figure has to be multiplied by 2, and actual figure for q = 2 is 98.8 meV The net effect of the distortion – symmetry lowering is the fermionic ground state energy stabilization.

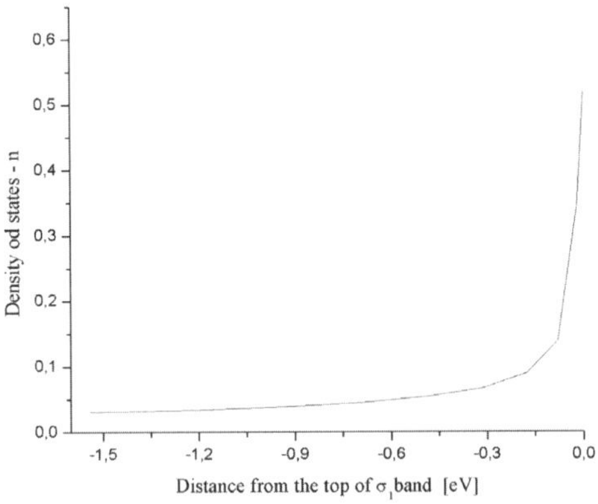

Figure II:2:5. Dependence of the σ_1 band density of states on the energy distance from the top of the band. For the relevant energy interval $\approx \hbar\omega_\Gamma$, the approximate mean value $n_{\sigma 1} = 0.2$ has been used. Outside of this interval, the density rapidly falls and reaches the value nearly ten times smaller, 0.03. For the upper σ band, the density of unoccupied states and occupied states at E_F are $n_{A(I)} = 0.02$. Density of unoccupied and occupied states of π band at E_F are calculated to be basically the same, $n_{B(J)} = 0.02$

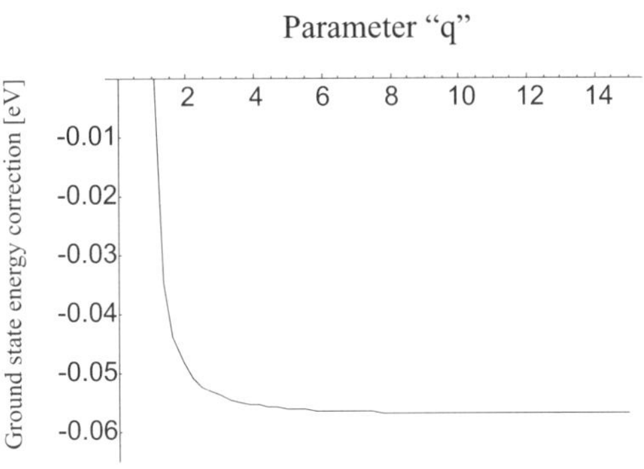

Figure II:2:6. The non-adiabatic correction to the fermionic ground state energy/unit cell of the MgB_2 as the function of the parameter "q" – see text. For q =2, it is –49.4 meV

It means that due to effective non-adiabatic EP coupling, the distorted structure for a displacement of $|\Delta f| \approx 0.0045$ - 0.005 (0.016 A°/atom) is by (-98,8 + 12) = - 86,8 meV more stable than undistorted – equilibrium structure on the BOA level. More over, this new – nonadiabatic ground state is geometrically degenerate. There are an infinite number of the B-B atoms in plane, out-of-phase displacements (see Fig.2), i.e. there can exist different nuclear configurations with the same ground state energy. On the lattice scale, geometrical degeneracy of the fermionic ground state energy for distorted structure, i.e. existence of an infinite number of out-of-phase B-B atoms displacements, enables cooperative and dissipationless motion of out-of-phase displaced B-B atoms along the perimeters of circles centered at the undistorted B-B atoms positions, with the same radii equal to the fraction displacement $|\Delta f| \approx 0.0045$ - 0.005 (≈ 0.016 A°). This is a new, coherent macroscopic quantum state.

2.4 Non-adiabatic correction to the one–particle term of the fermionic Hamiltonian. Corrections to the orbital energies and gap opening in a metallic one-particle spectrum.

For nonadiabatic correction, $\Delta_{\varepsilon I}$, to an occupied state ε_I^0 holds,

$$\Delta\varepsilon_I = \sum_A^{unocc} \Omega_{IA} - \sum_J^{occ} \Omega_{IJ} \tag{12}$$

In case of MgB_2, for quazicontinuum of states, with incorporation of the Lorenz line-shape form of the phonon mode, one can write:

$$\Delta\varepsilon_I = \frac{d}{\pi}\int_{-\infty}^{\infty} \frac{dx}{(x-\hbar\omega)^2 + d^2} \int_0^{\hbar\omega/q} n_B |U_{BI}|^2 \frac{x}{\left(\varepsilon_I^0 - \varepsilon_B^0\right)^2 - (x)^2} .d\varepsilon_B^0 \tag{13}$$

The original state ε_I^0 (orbital energy of the occupied state for particular clumped nuclear structure) is by this correction shifted on the energy scale to a new position,

$$\varepsilon_I = \varepsilon_I^0 + \Delta\varepsilon_I \tag{14}$$

For non-adiabatic correction, $\Delta\varepsilon_A$, to an unoccupied state ε_A^0 holds,

$$\Delta\varepsilon_A = \sum_B^{unocc} \Omega_{AB} - \sum_J^{occ} \Omega_{AJ}, \tag{15}$$

In case of MgB_2, for quazicontinuum of states, with incorporation of the Lorenz line-shape form of the phonon mode, holds :

$$\Delta\varepsilon_A = -\frac{d}{\pi}\int_{-\infty}^{\infty} \frac{dx}{(x-\hbar\omega)^2 + d^2} \int_{-\infty}^{-|\hbar\omega/q|} n_J |U_{AJ}|^2 \frac{x}{\left(\varepsilon_A^0 - \varepsilon_J^0\right)^2 - (x)^2} . d\varepsilon_J^0 \tag{16}$$

and the original state ε_A^0 (orbital energy of the unoccupied state for particular clumped nuclear structure) is shifted to a new position,

$$\varepsilon_A = \varepsilon_A^0 + \Delta\varepsilon_A \tag{17}$$

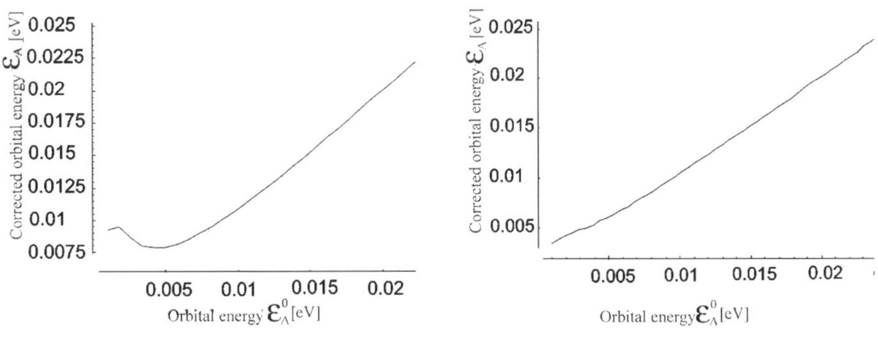

Figure II:2:7. The orbital energies of the unoccupied states after the non-adiabatic corrections (shifts) v.s. the original-uncorrected orbital energies of the σ_2 band (left) and π band (right). The minimum on the graphs indicates the energy position of the lowest unoccupied state of the σ_2 band (left) and π band (right) with respect to E_F (=0)

From the Figs.7. one can see that after the non-adiabatic corrections, the energy distance of the lowest unoccupied state from the E_F (half-gap) is 7.6 meV for σ_2 band and 4 meV for π band.
According to the gap equation [7],

$$\Delta(T) = \Delta(0).tgh\frac{\Delta(T)}{4k_BT} \tag{18}$$

calculated value of the critical temperature T_c is: $T_c(\sigma_2)$ = 39.5 K for σ_2 band gap, and $T_c(\pi)$ = 20.8 K for π band gap.

2.5 Non-adiabatic corrections to the density of states

For the final density of states that account for the effect of non-adiabaticity, the following equation can be derived [8]:

$$n(\varepsilon_Q) = \frac{1}{\left|1 + \frac{\partial(\Delta\varepsilon_Q)}{\partial\varepsilon_Q^0}\right|}.n(\varepsilon_Q^0) \quad , \tag{19}$$

with , $\Delta\varepsilon_Q = \varepsilon_Q - \varepsilon_Q^0$, and $n(\varepsilon_Q^0) = \left|\left(\frac{\partial\varepsilon_Q^0}{\partial k}\right)^{-1}\right|$

3. CONCLUSION

The non-adiabatic theory of EP interactions offers substantially different scenario of SC state transition than BCS one (or pairing theories in general). At stabilization of SC state, instead of Cooper pairs formation and Bose condensation at equilibrium geometry of clumped nuclei BOA structure, at the non-adiabatic level the crucial is symmetry breaking (nuclear displacements) resulting in a lower symmetry structure with the fermionic ground state energy geometrical degeneracy of atoms displacements and with the new one-particle spectrum that maximize non-adiabatic EP interactions. Meaning of the gap is also different here. In the non-adiabatic case, the gap has its usual meaning, i.e. quaziparticle ("non-adiabatic polaron") excitation energy over the one-particle spectrum.

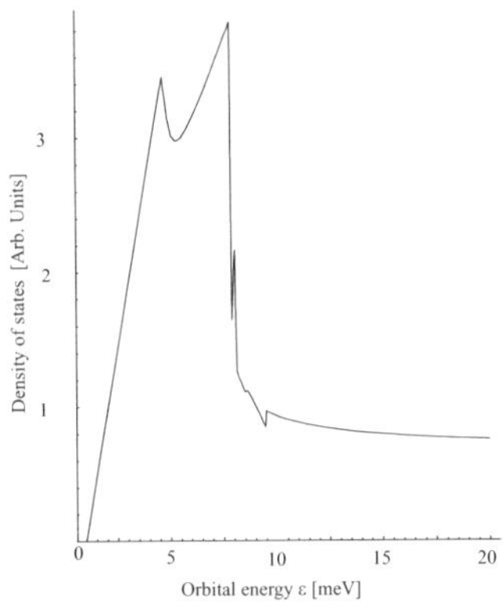

Figure II:2:8. The final – non-adiabatic density of the unoccupied states at 0 K, according to Eq.19.The uncorrected density of states is normalized to 1 and $E_F = 0$. The peak at 4 meV corresponds to non-adiabatic corrections of π band states and the peak at 7.6 meV corresponds to non-adiabatic corrections of σ_2 band states. The density of the occupied states is the mirror picture with respect to E_F

At finite temperatures, with increasing temperature from 0 K, due to Fermi statistics of the one-particle state populations, the non-adiabatic EP corrections become smaller (T–dependent contribution), and at crossing critical temperature T_c, the electronic energy loss due to symmetry lowering (T–independent contribution) becomes greater than the energy gain due to nonadiabatic EP interactions. The system, in order to minimize fermionic ground state energy, goes to normal – metal state with the "equilibrium" geometry of the higher symmetry. For cooling, the situation is opposite. At crossing T_c in downward direction, the distorted structure becomes more stable than the undistorted one. The reason is that at lowered symmetry, there is the proper structure of the one-particle spectrum at E_F , the structure that effectively enables to switch-on and maximizes the non-adiabatic EP interactions. The non-adiabatic energy gain starts to prevail over the energy loss of the distortion.

The SC state transition can be characterized as a non-adiabatic sudden increase of the cooperative kinetic effect at lattice energy stabilization (NASICKELES). It is exactly participation of the nuclear kinetic energy term on the non-adiabatic level that stabilize (negative contribution) fermionic ground state energy at a distorted structure. At the adiabatic conditions, nuclear kinetic effect is absent, and adiabatic correction (DBOC - effect of nuclear positions) to the fermionic ground state energy is always positive for equilibrium as well as for distorted clumped nuclei structures. It should be pointed out that for undistorted – equilibrium structure, the one-particle spectrum is of such character that non-adiabatic EP interactions do not contribute to the ground state energy stabilization. Only at the lowered symmetry – distorted structure, when the orbital energies are shifted into the "right" positions at E_F (formation of the new one-particle spectrum after distortion), the non-adiabatic EP interactions can become operative and effective. Geometrical degeneracy of the fermionic ground state energy for distorted structure, i.e. existence of an infinite number of out of phase B-B atoms displacements, enables cooperative motion (cooperative nuclear microcirculations) of out-of-phase displaced B-B atoms along the perimeters of circles centered at the undistorted B-B atoms positions, with the same radii equal to the fraction displacement $|\Delta f|$.

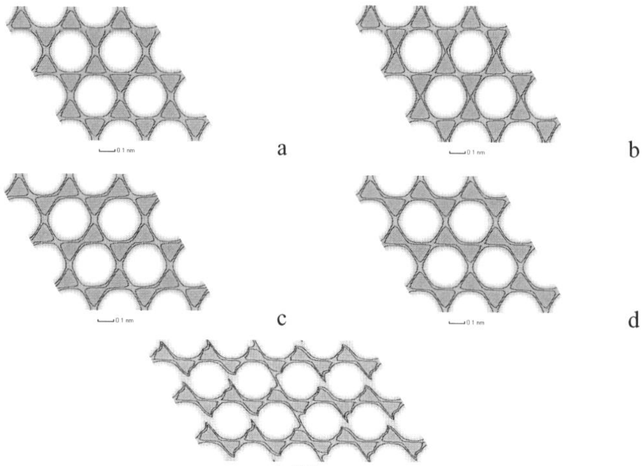

*Figure II:2:9*a-e. The valence electron iso-density lines in the plane of B atoms (**a-b** plane) for equilibrium (a) and distorted structures (b-e). The electron density is localized at B atom positions for equilibrium structure (a). The B atoms displacements ($|\Delta f| = 0.005$) induce the alternating interatomic charge density delocalization, different for the particular types of the distortion (b-d). Nuclear "microcirculation" enables then effective charge transfer over the lattice in an external electric potential. The Fig (e) corresponds to the case of the distortion (d) over the larger lattice segment

In the present case, the radius of circles is $|\Delta f| = 0.005$ (expressed in fraction unit), that is 0.016 A° in the absolute value. The cooperative nuclear motion, i.e. "nuclear microcirculations", induces dynamic - cooperative formation of shortened and elongated in-plane B-B bond distances on the lattice scale, with a dynamic formation of increased and decreased interatomic charge densities – see Fig 9a-e, i.e. dynamic formation of non-adiabatic bipolarons.

REFERENCES

1. M.Svrček, P. Baňacký, A. Zajac, Int. J. Quant. Chem. 43, 393 (1992).
2. M. Svrček, P. Baňacký, A. Zajac, Int. J. Quant. Chem. 43, 551 (1992).
3. M. Svrček, P. Baňacký, A. Zajac, Int. J. Quant. Chem. 43, 425 (1992).
4. J. Noga, P. Baňacký, S. Biskupič, et al., J. Comp. Chem. 20, 253 (1999).
5. SOLID 2000, S-Tech a.s., Bratislava, Slovakia (www.stech.sk).
6. T. Yildirim, O. Gulseren, J.W. Lynn, et al., Cond-mat/0103469.
7. M. Svrček, P. Baňacký, A. Zajac, Int. J. Quant. Chem. 43, 415 (1992).
8. P. Baňacký, M. Svrček, V.Szocs, Int. J. Quant. Chem. 58, 487 (1996).

III

PHASE SEPARATION AND TWO COMPONENTS CUPRATES

III.1

MICROSCOPIC PHASE SEPARATION AND TWO TYPE OF QUASIPARTICLES IN LIGHTLY DOPED $La_{2-x}Sr_xCuO_4$ OBSERVED BY ELECTRON PARAMAGNETIC RESONANCE

A. Shengelaya,[1] M. Bruun,[1] B. I. Kochelaev,[2] A. Safina,[2] K. Conder,[3] and K. A. Müller[1]

[1]Physik-Institut der Universität Zürich, Winterthurerstrasse 190, CH-8057, Switzerland
[2]Department of Physics, Kazan State University, Kazan, 420008, Russia
[3]Laboratory for Neutron Scattering, ETH Zürich and PSI, CH-5232 Villigen PSI, Switzerland

Abstract: In the low doping range of x from 0.01 to 0.06 in $La_{2-x}Sr_xCuO_4$, we observed two electron paramagnetic resonance (EPR) signals: a narrow and a broad one. The narrow line is ascribed to metallic regions in the material, and its intensity increases exponentially upon cooling below ~ 150 K. The activation energy deduced $\Delta = 460(50)$ K is nearly the same as that found in the doped superconducting regime by Raman and neutron scattering. Obtained results provide evidence of the microscopic phase separation and two type of quasiparticles in lightly doped $La_{2-x}Sr_xCuO_4$

PACS numbers: 74.25.Dw, 74.72.Dn, 74.20.Mn, 76.30.-v

The mechanism of high-T_c superconductivity (HTSC) remains enigmatic even after 17 years of its discovery [1]. It is known that HTSC is achieved when a moderate density of conducting holes is introduced in the CuO_2 planes. At a critical concentration of doping $x_{cr} \approx 0.06$, superconductivity appears. However, it is still an unresolved issue how the electronic structure evolves with hole doping from the antiferromagnetic insulator to the paramagnetic metalllic and superconducting state. While most of the

A. Bianconi (ed.), Symmetry and Heterogeneity in High Temperature Superconductors, 105–116.

theoretical models assume that the holes are homogeneously doped into CuO_2 planes, there are increasing number of the experiments pointing towards highly nonuniform hole distribution leading to a formation of hole-rich and hole-poor regions [2]. This electronic phase separation is expected to be mostly pronounced at low hole concentrations. In the early experiments Johnston *et al.* analized the susceptibility of the $La_{2-x}Sr_xCuO_4$ (LSCO) samples at concentrations $x < x_{cr}$ using the finite-size scaling and concluded that the material consists of antiferromagnetic (AF) domains of variable size, separated by metallic domain walls [3]. More recently Ando *et al.* corroborated this early finding by measuring the in-plane resistivity anisotropy in untwinned single crystals of $La_{2-x}Sr_xCuO_4$ (LSCO) and $YBa_2Cu_3O_{7-\delta}$ (YBCO) in the lightly doped region, interpreting their results in terms of metallic stripes present [4]. Furthermore, most recently, magnetic axis rotation was reported, and points to a high mobility of the crystallographic (metallic) domain boudaries of the AFM domains [5]. The Coulomb interaction limits the spatial extention of the electronic phase separation to hole-rich and hole-poor regions to a microscopic scale [6]. Therefore it is important to use a local microscopic methods to study the electronic phase separation in cuprates.

The nuclear magnetic resonance (NMR) is one of such methods. However, in lightly doped cuprates with $x < x_{cr}$ the NMR signal from Cu wipes out due to the strong AF fluctuations and no NMR signal can be observed at low temperatures [7]. On the other hand an electron paramagnetic resonance (EPR) signal can still be observed because the time-domain of observation of EPR is two to three orders of magnitude shorter than that of NMR [8]. In order to observe an EPR signal we doped LSCO with 2% of Mn, which in the 2+ valent state gives a well defined signal and substitutes for the Cu^{2+} in the CuO_2 plane. It was shown by Kochelaev et al. that the Mn ions are strongly coupled to the collective motion of the Cu spins (the so called bottleneck regime) [9]. Recently we have studied the EPR of Mn^{2+} in $La_{2-x}Sr_xCuO_4$ in a doping range $0.06 \leq x \leq 0.20$ [8]. The bottleneck regime allowed to obtain substantial information on the dynamics of the copper electron spins in the CuO_2 plane as a function of Sr doping and oxygen isotope substitution. In the present work we performed a thorough EPR investigation of the LSCO in lightly doped range $0.01 \leq x \leq 0.06$, i.e. below x_{cr}.

The $La_{2-x}Sr_xCu_{0.98}Mn_{0.02}O_4$ polycrystalline samples with $0 \leq x \leq 0.06$ were prepared by the standard solid-state reaction method. The EPR measurements were performed at 9.4 GHz using a BRUKER ER-200D spectrometer equipped with an Oxford Instruments helium flow cryostat. In order to avoid a signal distortion due to skin effects, the samples were ground and the powder was suspended in paraffin. We observed an EPR signal in all samples. The signal is centered near $g \sim 2$, a value very close to the g-factor for the Mn^{2+} ion.

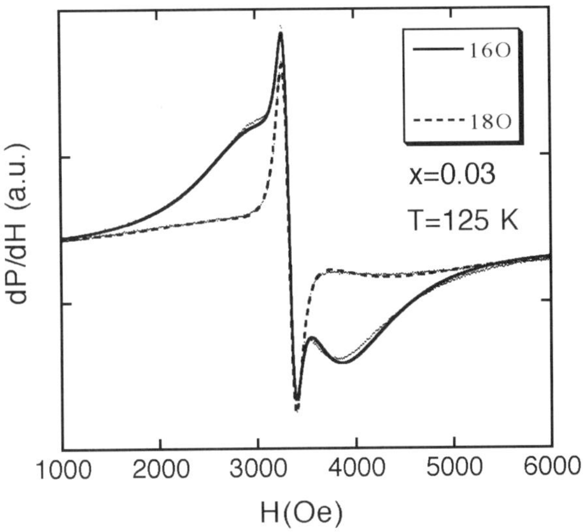

Figure III:1:1. EPR signal of ^{16}O and ^{18}O samples of $La_{1.97}Sr_{0.03}Cu_{0.98}Mn_{0.02}O_4$ measured at T=125 K under identical experimental conditions. The solid and dashed lines represent the best fits using a sum of two Lorentzian components with different linewidths: a narrow and a broad one

Figure 1 shows the EPR lines observed in x=0.03 sample with different oxygen isotopes ^{16}O and ^{18}O. First, one should note that the EPR spectra consist of two lines. We found that they can be well fitted by a sum of two Lorentzian components with different line widths: a narrow and a broad one. The second important observation is that the narrow line shows practically no isotope effect, whereas the broad line exhibits a huge isotope effect. Similar two-component EPR spectra were observed in other samples with different Sr concentrations up to $x = 0.06$. At $x = 0.06$, only a single EPR line is seen in the entire temperature range, in agreement with our previous studies of samples with $0.06 \leq x \leq 0.20$ [8].

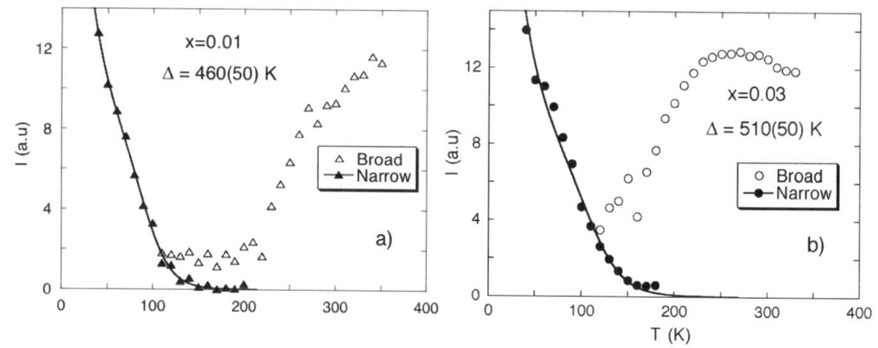

Figure III:1:2. Temperature dependence of the narrow and broad EPR signal intensity in $La_{2-x}SrxCu_{0:98}Mn_{0:02}O4$ with different Sr dopings: (a) x = 0.01; (b) x = 0.03. The solid lines represent fits using the model described in the text

Figure 2 shows the temperature dependence of the EPR intensity for samples with different Sr concentrations. One can see that the two EPR lines follow a completely different temperature dependence. The intensity of the broad line has a maximum and strongly decreases with decreasing temperature. On the other hand, the intensity of the narrow line is negligible at high temperatures and starts to increase almost exponentially below ~150 K. We note that the temperature below which the intensity of the broad line decreases shifts to lower temperatures with increasing doping. However, the shape of the $I(T)$ dependence for the *narrow line is practically doping-independent* and only slightly shifts towards higher temperatures with increased doping. A similar tendency is observed also for the temperature de pendence of the EPR linewidth. The linewidth of the broad line and its temperature dependence are strongly doping-dependent, whereas the linewidth of the narrow line is very similar for samples with different Sr doping (see Fig. 3).

It is important to point out that the observed two-component EPR spectra are an intrinsic property of the lightly doped LSCO and are not due to conventional chemical phase separation. We examined our samples using x-ray diffraction, and detected no impurity phases. Moreover, the temperature dependence of the relative intensities of the two EPR signals rules out macroscopic inhomogeneities and points towards a microscopic electronic phase separation. The qualitatively different behavior of the broad and narrow EPR signals indicates that they belong to distinct regions in the sample. First we notice that the broad line vanishes at low temperatures. This can be explained by taking into account the AF order present in samples with very low Sr concentration [3]. It is expected that upon approaching the AF ordering temperature, a strong shift of the resonance frequency and an increase of the relaxation rate of the Cu spin system will occur. This will break the bottleneck regime of the Mn^{2+} ions, and as a consequence the EPR signal becomes unobservable [8].

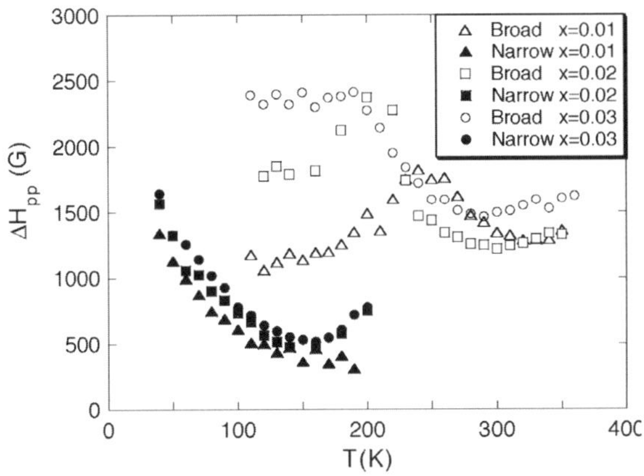

Figure III:1:3. Temperature dependence of the peak-to-peak linewidth ΔHpp for the narrow and broad EPR lines in $La_{2-x}Sr_xCu_{0:98}Mn_{0:02}O_4$ with x=0.01, 0.02 and 0.03

In contrast to the broad line, the narrow signal appears at low temperatures and its intensity increases with decreasing temperature. This indicates that the narrow signal is due to the regions where the AF order is supressed. It is known that the AF order is destroyed by the doped holes, and above $x = 0.06$ AF fluctuations are much less pronounced [10]. Therefore, it is natural to relate the narrow line to regions with locally high carrier concentration and high mobility. This assumption is strongly supported by the absence of an oxygen isotope effect on the linewidth of the narrow line as well. It was shown previously that the isotope effect on the linewidth decreases at high charge-carrier concentrations close to the optimum doping [8]. We obtain another important indication from the temperature dependence of the EPR intensity. Because we relate the narrow line to hole-rich regions, an exponential increase of its intensity at low temperatures indicates an energy gap for the existence of these regions. In the following we will argue that this phase separation is assisted by the electron-phonon coupling. More precisely, the latter induces anisotropic interactions between the holes via the phonon exchange, resulting in the creation of extended nano-scale hole-rich regions.

An interaction between holes via the phonon exchange can be written in the form [11]:

$$H_{pol-ph} = G\sum_{n\alpha} P^{+}_{n\alpha} P_{n\alpha} \in_{\alpha\alpha} \tag{1}$$

where $P^{+}_{n\alpha}$ is a creation operator of one polaron, $\in_{\alpha\alpha}$ is a deformation tensor, G is a coupling constant. It was shown that this interaction reduces to usual elastic forces if we neglect the retardation effects and optical modes. Following Aminov and Kochelaev [11], Orbach and Tachiki [12] we can find an interaction due to an exchange by phonons between two holes oriented along the axes α and β separated by the space vector $R = R_{n\alpha} - R_{m\beta}$

$$H_{\text{int}} = F(R_{n\alpha} - R_{m\beta}) P^{+}_{n\alpha} P_{n\alpha} P^{+}_{m\beta} P_{m\beta}; \tag{2}$$

$$F_{xx}(\mathrm{R}) = \frac{1}{8\pi\rho C_l^2} \frac{G^2}{R^3} \left\{ 2\left(1 - 3\gamma_x^2\right) + \eta\left[12\gamma_x^2 - 15\gamma_x^4 - 1\right] \right\},$$

$$F_{xy}(\mathrm{R}) = \frac{1}{8\pi\rho C_l^2} \frac{G^2}{R^3} \left(2 - 15\gamma_x^2\gamma_y^2\right).$$

Here C_l, C_t are longitudinal and transversal sound velocities; $\gamma_x = x/R$, $\gamma_y = y/R$; $\eta = (C_l^2 - C_t^2)/C_t^2$. This interaction is highly anisotropic being attractive for some orientations and repulsive for others [13]. The attraction between holes may result in a bipolaron formation when holes approach each other closely enough. The bipolaron formation can be a starting point for the creation of hole-rich regions by attracting of additional holes. Because of the highly anisotropic elastic forces these regions are expected to have the form of stripes. Therefore the bipolaron formation energy Δ can be considered as an energy gap for the formation of hole-rich regions.

In applying the above model to the interpretation of our EPR results we have to keep in mind that the spin dynamics of the coupled Mn-Cu system experiences a strong bottleneck regime [9]. In the bottleneck regime the collective motion of the total magnetic moment of the Mn and Cu spin system appears because the relaxation rate between the magnetic moments of the Mn and Cu ions due to the strong isotropic Mn-Cu exchange interaction is much faster then their relaxation rates to the lattice. The intensity of the joint EPR signal, being proportional to the sum of spin susceptibilities $I \sim \chi_{Mn} + \chi_{Cu}$, is determined mainly by the Mn susceptibility, since $\chi_{Mn} \gg \chi_{Cu}$ for our Mn concentration and temperature range. This results in a Curie-Weiss temperature dependence of the EPR signal.

Taking into account this remark we conclude that the EPR intensity of the narrow line is proportional to the volume of the sample occuppied by the hole-rich regions because the Mn ions are randomly distributed in the sample. We expect that the volume in question is proportional to the number of bipolarons, which can be estimated in a way proposed by Mihailovic and Kabanov [14]. If the density of states is determined by $N(E) \sim E^{\alpha}$, the number of bipolarons is

$$N_{bipol} = \left(\sqrt{z^2 + x - z}\right)^2, \quad z = KT^{\alpha+1} exp\left(-\frac{\Delta}{T}\right) \tag{3}$$

where Δ is the bipolaron formation energy, x is the level of hole doping, and K is a temperature- and doping-independent parameter related to the free polaron density of states. The EPR intensity from the hole-rich regions will be proportional to the product of the Curie-Weiss susceptibility of the bottlenecked Mn-Cu system and the number of the bipolarons

$$I_{\text{narrow}} \sim \frac{C}{T - \theta} N_{\text{bipol}} \,. \tag{4}$$

where C is the Curie constant and θ is the Curie-Weiss temperature. The experimental points for the narrow-line intensity were fitted for the two-dimensional system (α=0), and we used the value θ = -8 K, which was found from measurements of the static magnetic susuceptibility (an attempt to vary θ yielded about the same value). The values of C and θ are determined mainly by the concentration and magnetic moment of the Mn ions and their coupling with the Cu ions. Since these parameters are expected to be doping independent (or weakly dependent), they were found by fitting for the sample x=0.01 and then were kept constant for other concentrations leaving the only free parameter the energy gap Δ.

The results of the fit are shown in Fig. 2 (a,b). For the bipolaron formation energy we obtained Δ =460(50) K, which is practically doping-independent. This value agrees very well with the value of Δ obtained from the analysis of inelastic neutron-scattering and Raman data in cuprate superconductors [14]. Recently Kochelaev et al. performed theoretical calculations of the polaron interactions via the phonon field using the extended Hubbard model [13]. They estimated the bipolaron formation energy and obtained values of $100 \text{ K} \leq \Delta \leq 730 \text{ K}$, depending on the value of the Coulomb repulsion between holes on neighboring copper and oxygen sites V_{pd}, $0 \leq V_{pd} \leq 1.2$ eV. This means that the experimental value of Δ can be understood in terms of the elastic interactions between the polarons.

It is interesting to compare our results with other experiments performed in lightly doped LSCO. Recently Ando *et al.* measured the in-plane anisotropy of the resistivity ρ_b/ρ_a in single crystals of LSCO with x = 0.02-0.04 [4]. They found that at high temperatures the anisotropy is small, which is consistent with the weak orthorhombicity present. However, ρ_b/ρ_a grows rapidly with decreasing temperature below ~ 150 K. This provides macroscopic evidence that electrons self-organize into an anisotropic state because there is no other external source to cause the in-plane anisotropy in $La_{2-x}Sr_xCuO_4$. We notice that the temperature dependence of the narrow EPR line intensity is very similar to that of ρ_b/ρ_a obtained by Ando *et al.* (see Fig. 2(d) in Ref. 4). To make this similarity clear, we plotted $I_{narrow}(T)$ and ρ_b/ρ_a (T) on the same graph (see Fig. 4).

It is remarkable that both quantities show very similar temperature dependences. It means that our *microscopic* EPR measurements and the *macroscopic* resistivity measurements by Ando *et al.* provide evidence of the same phenomenon: the formation of hole-rich metallic stripes in lightly doped LSCO well below x_{cr} = 0.06. This conclusion is also supported by a recent angle-resolved photoemission (ARPES) study of LSCO which clearly demonstrated that the metallic quasiparticles exist near the nodal direction below x=0.06 [16].

A number of experiments on HTSC suggest the possible existence of two quasiparticles: a heavy polaron and a light fermion [15]. In the context of the two-carrier paradigm, the narrow line in the EPR spectra may be attributed to centers with nearly undistorted environment in regions where carriers are highly mobile, wherease the broad line is due to the centers with a distorted environment and slow polaronic carriers.

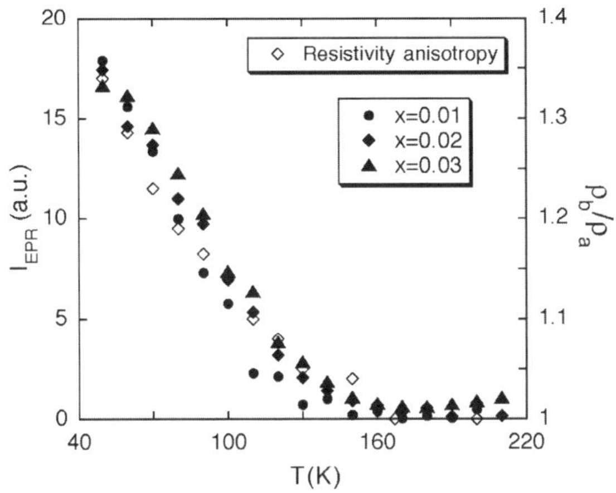

Figure III:1:4. Temperature dependence of the narrow EPR line intensities in $La_{2-x}Sr_xCu_{0:98}Mn_{0:02}O_4$ and of the resistivity anisotropy ratio in $La_{1:97}Sr_{0:03}CuO_4$ obtained in Ref. 4

Recent high-resolution ARPES experiments by Lanzara *et al.* clearly showed a kink in the quasiparticle energy versus wavevector plots for different HTSC [17]. The kink at an energy of about 70 meV separates the dispersion into a high-energy part (that is, further from the Fermi energy) and a low-energy part (that is, closer to the Fermi energy) with different slopes. The two different group velocities above and below the kink are probably due to two quasiparticles with different effective masses. In this case a low-energy part of the dispersion would correspond to our narrow EPR line and the high-energy part to broad EPR line. This assumption is strongly supported by very recent isotope effect ARPES experiments by Lanzara *et al.* [18]. They observed that the high-energy quasiparticles have a strong isotope effect, while the low-energy quasiparticles show practically no isotope effect. This is in direct correspondence with our EPR results where the broad EPR line due to heavy polaronic quasiparticles shows a huge isotope effect, whereas the narrow line stemming from metallic regions has no isotope effect. Moreover, looking at Figs. 2,3 one can see that the intensity and linewidth of the narrow line is practically doping independent, while both the intensity and linewidth of broad line change strongly with the doping. This shows that the local electronic properties of the metallic regions which appear due to the microscopic phase separation and yeld the narrow EPR line are the same in different samples despite different doping levels. This is again in excellent agreement with the ARPES data showing that the Fermi velocity of low-energy quasiparticles are independent of doping in different cuprate families, while the high-energy velocity varies strongly with doping [19]. Also, transport measurements showed that the mobility of charge carriers in LSCO at moderate temperatures remains the same throughout a wide doping range from the lightly-doped AF to optimally doped superconducting regime [20]. These results suggest that the electronic transport is governed by essentially the same mechanism from lightly doped to optimally doped range.

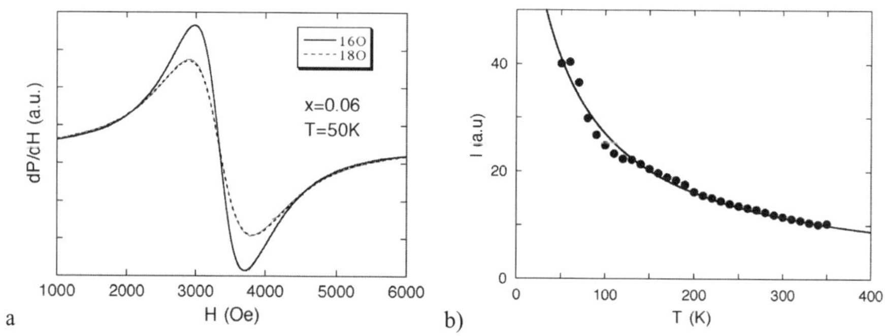

Figure III:1:5. (a): EPR signal of ^{16}O and ^{18}O samples of $La_{1:94}Sr_{0:06}Cu_{0:98}Mn_{0:02}O4$ measured at T=50 K under identical experimental conditions. The solid and dashed lines represent the best fits using a single Lorentzian. (b):Temperature dependence of the EPR signal intensity in $La_{1:94}Sr_{0:06}Cu_{0:98}Mn_{0:02}O_4$. The solid line represents the fit using the Curie-Weiss temperature dependence with the Curie-Weiss temperature θ=40 K figure]

Figure 5 shows the EPR spectra and intensity for the x=0.06 sample. It is seen that in contrast to the x <0.06 samples at x=0.06 only a single EPR line is observed and the temperature dependence of the signal intensity recovers an usual Curie-Weiss behavior. On the other hand there is still a substantial isotope effect on the EPR line. To understand the change of the EPR spectra at x=0.06, one should first comment on the observability of the phase separation in our EPR experiments. The main difference of the EPR signals from the hole-rich and hole-poor regions is the spin relaxation rate of the Cu spin system, which results in different EPR linewidths. One would expect these local differences of the relaxation rate to be averaged out by the spin diffusion. The spin diffusion in the CuO_2 plane is expected to be very fast because of the huge exchange integral between the Cu ions. A rough estimate shows that during the Larmor period a local spin temperature can be transported over 100 Cu-Cu distances. It means that all the different nanoscale regions will relax to the lattice with a single relaxation rate, and we cannot distinguish them with EPR. However, the AF order which appears below T_N in the hole-poor regions in lightly doped LSCO freezes the process of spin diffusion, and this is the reason we can see different EPR lines from the two types of regions. From this we expect that with increasing doping, where magnetic order gets suppressed, spin diffusion will become faster, extended, and we can no longer distinguish different regions with EPR. This is most probably what happens in samples with $x \geq 0.06$, where only a single EPR line is observed [8]. This does not mean that the phase separation in hole-rich and hole-poor regions does not exist at $x \geq 0.06$, but that the spin diffusion averages out the EPR response from these regions. In fact, ARPES measurements showed the presence of two quasiparticles in the whole doping range [17, 19], indicating that the electronic phase separation exists also at higher doping levels. Also, recent Raman and infrared measurements provided evidence of one-dimensional conductivity in LSCO with $x = 0.10$ [21].

In summary, EPR measurements in lightly doped LSCO revealed the presense of two resonance signals: a narrow and a broad one. Their behavior indicates that the narrow signal is due to hole-rich metallic regions and the broad signal due to hole-poor AF regions.The narrow-line intensity is small at high temperatures and increases exponentially below ~ 150 K. The activation energy inferred, Δ= 460(50) K, is nearly the same as that deduced from other experiments for the formation of bipolarons, pointing to the origin of the metallic stripes present. We found a remarkable similarity between the temperature dependences of the narrow-line intensity and recently measured resistivity anisotropy in CuO_2 planes in lightly doped LSCO [4]. The results obtained provide the first magnetic resonance evidence of the formation of hole-rich metallic domains in lightly doped LSCO well below $x_{c1} = 0.06$. These results also lend support to the recently proposed model of phase coherence by percolation (PSP) [22]. According to the PSP model the macroscopic phase coherence in superconducting cuprates is determined by random percolation between mesoscopic bipolarons, stripes or clusters. This simple model allows to quantitatively predict the critical carrier concentration for the occurence of the superconductivity as well as to estimate T_c from experimentally measured values of the pairing energy.

This work is supported by SNSF under the grant 7IP062595, BIK and AS are partially supported by INTAS-01-0654 and CRDF REC-007.

REFERENCES

1. J. G. Bednorz and K. A. Müller, Z. Phys. B 64, 189 (1986).
2. See, for instance, *Phase separation in Cuprate Superconductors*, edited by K.A. Müller and G. Benedek (World Scientific, Singapore, 1993).
3. J. H. Cho *et al.*, Phys. Rev. Lett. 70, 222 (1993).
4. Y. Ando *et al.*, Phys. Rev. Lett. 88, 137005 (2002).
5. A.N. Lavrov, S. Komiya and Y. Ando, Nature 418, 385 (2002).
6. V. J. Emery and S. A. Kivelson *et al.*, Physica C 209, 597 (1993).
7. T. Imai *et al.*, Phys. Rev. Lett. 70, 1002 (1993).
8. A. Shengelaya *et al.*, Phys. Rev. B 63, 144513 (2001).
9. B. I. Kochelaev *et al.*, Phys. Rev. B 49, 13106 (1994).
10. Ch. Niedermayer *et al.*, Phys. Rev. Lett. 80, 3843 (1998).
11. L. K. Aminov and B. I. Kochelaev, Zh. Eksp. Theor. Fiz. 42, 1303 (1962).
12. R. Orbach and M. Tachiki, Phys. Rev. 158, 524 (1967).
13. B. I. Kochelaev *et al.*, Mod. Phys. Lett. B 17, 415 (2003).
14. V. V. Kabanov and D. Mihailovic, Phys. Rev. B65, 212508 (2002).
15. D. Mihailovic and K. A. Müller, in *High-Tc Superconductivity: Ten Years After the Discovery*, E. Kaldis *et al.*, eds. Kluwer Academic Publishers (1997) p. 243.
16. T. Yoshida *et al.*, Phys. Rev. Lett. 91, 027001 (2003).
17. A. Lanzara *et al.*, Nature 412, 510 (2001).
18. A. Lanzara *et al.*, unpublished.
19. X. J. Zhou *et al.*, Nature 423, 398 (2003).
20. Y. Ando *et al.*, Phys. Rev. Lett. 87, 017001 (2001).
21. F. Venturini *et al.*, Phys. Rev. B 66, R060502 (2002).
22. D. Mihailovic *et al.*, Europhys. Lett. 57, 254 (2002).

III.2

PHASE SEPARATION IN CUPRATES INDUCED BY DOPING, HYDROSTATIC PRESSURE OR ATOMIC SUBSTITUTION

E. Liarokapis
Department of Physics, National Technical University, 15780 Athens, Greece

Abstract: Results of systematic Raman studies of phase separation induced on cuprates by doping, hydrostatic pressure, and atomic substitutions are presented. In $YBa_2Cu_3O_x$, oxygen doping induces a separation into microphases from the ordering of the chains, at relative amounts that vary with concentration. In the overdoped region, a similar coexistence of phases appears, which is due to the modification of the CuO_2 buckling. The excess doping of $YBa_2Cu_3O_{7-\delta}$ by Ca creates domains of pure Y123 and others where the Ca atoms are surrounded by yttrium as first neighbours. Hydrostatic pressure induces non-linear effects in almost all A_g–symmetry phonons of $YBa_2Cu_3O_x$, $YBa_2Cu_4O_8$, and the $Bi_2Sr_2CaCu_2O_8$ superconductors. A coexistence of phases seems to appear at a critical pressure where changes in the superconducting transition temperature have been observed.

1. INTRODUCTION

Structural and electronic inhomogeneities are characteristic intrinsic properties of the cuprate superconductors related with the unconventional character of these compounds. The $YBa_2Cu_3O_x$ (Y123x) compound is a prototype of cuprates with two CuO_2 planes been extensively investigated over the $0 \leq x \leq 7.0$ range of oxygen concentration [1]. The properties and the crystal structure of this material are closely related to x, with

A. Bianconi (ed.), Symmetry and Heterogeneity in High Temperature Superconductors, 117–132.

superconductivity appearing for $x \geq 6.4$. Ordering of the chains results in the formation of various phases, such as the ortho-II phase having a transition temperature of 60K, and the optimally doped phase with T_c=92K for x=6.92. On the other hand, oxygen concentrations above nominal optimal amounts (x=7) are not easily attained, and overdoping of Y1237 is usually achieved by Ca substitution for Y. This substitution was originally considered to follow the generally accepted scheme of the increase of carriers [2]. Raman spectroscopy is suitable for investigating the local structural order, due to its discrimination limit of a few unit cells. Results on Raman scattering of Ca doped Y1237 will be examined in order to investigate the role of the additional carriers for the reduction of the T_c.

In Y123x a systematic Raman study has discovered an anomalous softening of the (A_g symmetry) in-phase vibrations along the c-axis of the oxygen atoms O_{pl} of the CuO_2 planes at the optimal to overdoped oxygen concentrations [3]. This decrease in energy with the addition of carriers beyond x~6.92 doping is accompanied with a reduction in T_c [4]. Actually a gap has been observed for x>6.95 in the distribution of the energy of this phonon, pointing out to the coexistence of phases in the overdoped region (x>6.975) and to a 1st order phase transition [3]. This behaviour was found to be independent of the temperature [5]. Magnetic measurements have shown that there are two transition temperatures above optimal doping pointing again to a coexistence of phases [6]. Finally, EXAFS measurements have shown that there is a local structural modification at the optimal to overdoped region, which originates from a sudden change of the buckling of the CuO_2 planes [7].

A more clear way than chemical doping to investigate the effects of the local structural modifications and the phase separation to the transition temperature is the application of a hydrostatic pressure. The effect of the hydrostatic pressure in most of the cuprates has been studied extensively [8,9]. Usually, the dependence of the superconducting transition temperature on hydrostatic pressure (dT_c/dP) varies in a large range of values, indicating in some cases the tendency of the material to reach higher T_c values. Single crystals of $YBa_2Cu_3O_x$ (with x=6.5 and 7), $YBa_2Cu_4O_8$, and $Bi_2Sr_2CaCu_2O_8$ have been studied under hydrostatic pressure by Raman spectroscopy [10-13]. In all cases the results show a clear correlation between the spectral modifications and the transition temperature. Whenever structural data at high pressures are available, they also indicate local structural modifications at the characteristic pressures where T_c is modified.

2. EXPERIMENTAL

Raman scattering results were obtained from several polycrystalline compounds of $YBa_2Cu_3O_x$ (Y123:x, with 6<x<7) and $Y_{1-x}Ca_xBa_2Cu_3O_{7-\delta}$ (Y-Ca123) and from single crystals of $Bi_2Sr_2CaCu_2O_8$ (Bi2212) and $YBa_2Cu_4O_8$ (Y124). All measurements have been carried out at room and low temperatures and at high hydrostatic pressures using a T64000 Jobin-Yvon triple spectrometer. A liquid nitrogen cooled charge coupled device (CCD) was the detector and the spectrometer was equipped with a microscope of magnification ×100 (×40 for the hydrostatic pressure measurements). For the high-pressure measurements (up to 7.5 GPa) a Merrill-Bassett type diamond anvil cell (DAC) was used, which has allowed the Raman studies to be carried out in a back scattering geometry. The pressure-transmitting medium was a mixture of methanol-ethanol. For monitoring the pressure a small "standard" silicon crystal was used inside the gasket. The beam from several lines of an Ar^+ laser was focused on the sample at a spot of diameter 1-2 μm (100) or 3-5 μm (40), while the power level was kept below 0.20 mW (or 0.40 mW for the high pressure measurements). Typical accumulation times were 1-2 hours (room temperature) or 2-6 hours (high pressure) depending on the scattering polarization. Always freshly cut samples were used for the study of the various scattering geometries.

3. RESULTS AND DISCUSSION

3.1 Doping induced phase separation

The formation of phases in $YBa_2Cu_3O_x$ is mostly related with the chain ordering [1]. In the Raman spectra, characteristic changes are observed in the phonons due to the apical oxygen O_{ap} and the in-phase vibrations of the plane oxygen atoms O_p related with the formation of the ortho-II, ortho-III, and the tetragonal phases [1,4,5]. With increasing amount of oxygen, the peak position of the apical oxygen atom shifts to higher energies roughly as the T_c changes with doping (Fig.1). Its width is a non-linear function of doping with minima at $x \geq 6.94$, $x \cong 6.45$-6.47, and $x \leq 6.15$ (Fig.1). Both the energy and the width of the in-phase phonon remain constant in the insulating range (Fig.2). The mode energy decreases and the width increases with oxygen doping for x>6.3. These changes can be attributed to the formation of phases [1,4,5] from the ordering of the chain oxygen atoms.

In Fig.2 one can see another spectral modification of the in-phase phonon, which for $x \geq x_{opt}$ softens by ~7 cm^{-1}.

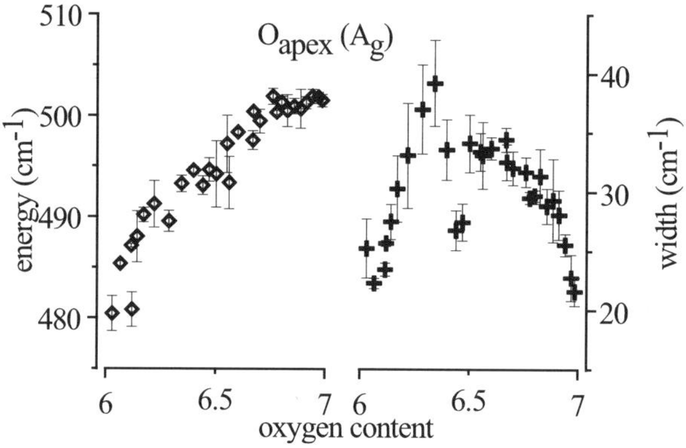

Figure III:2:1. The energy (left) and width (right) of the A_g-symmetry apical oxygen atom

The amount of softening is independent of temperature (the same above or below T_c) [5]. Magnetic measurements have shown that there are two transition temperatures, one related with the optimal doping at 92K and another of the overdoped phase with a reduced transition temperature of 88K [6]. At the optimal doping an anomaly on the c-axis has been discovered by neutron scattering and XRD (for samples prepared with certain methodology), both related with a sudden increase of the Cu_{pl}-O_{pl} distances [1]. The use of EXAFS as a local probe at the neighbor of the Y atom has shown that in all samples there is a sudden increase of the buckling at the phase transition from the optimal to overdoping [7].

Besides, in the Raman spectra one could detect in different microcrystallites either the high energy value of the in-phase phonon, which corresponds to the optimal doping or to the lower by ~7 cm^{-1}, apparently linked to the overdoped phase with the reduced transition temperature and the increased buckling angle [1]. All these results from various methodologies point to a spatial inhomogeneity driven by doping with coexisting phases like a 1st order phase transition.

The softening of the in-phase phonon in the overdoped region can be attributed to a double well potential and a strong anharmonicity of this phonon. Systematic Raman measurements of the oxygen anharmonicity of $YBa_2Cu_3O_x$ have been carried out in the whole range of doping [14]. The results have shown that the B_{1g} phonon of the O_{pl} does not show any anharmonicity in the whole range of doping [14]. For the apex mode there is a slight deviation in the oxygen concentration region 6.4<x<7.0, which diminishes towards x~6.4 and x~6.9 and this can be attributed to the coexistence of phases from the chain ordering [14].

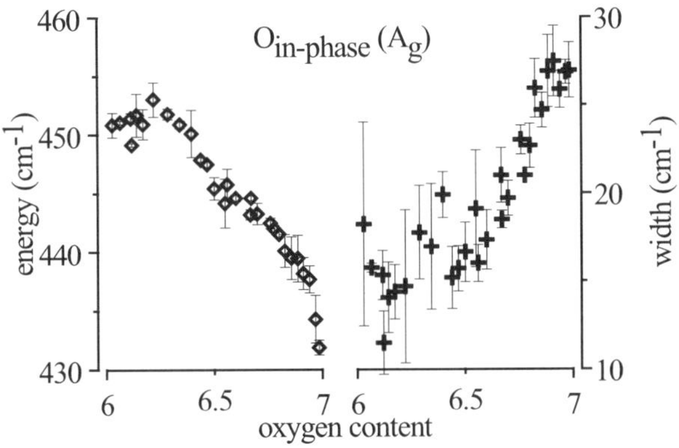

Figure III:2:2. Energy and width of in-phase phonon

In Fig.3 the energy of the in-phase mode is normalized to the B_{1g} phonon energy in order to cancel uncertainties in the amount of ^{16}O substituted by ^{18}O. It is clear that the in-phase phonon shows a decreasing amount of anharmonicity for x>6.5 and the anharmonicity disappears completely at optimal doping. On the other hand, there is anharmonicity at lower oxygen concentrations, which may originate from the coexistence of chain-ordering phases

The buckling of the CuO_2 is mostly induced by the unequal charge of the Ba and the Y atoms in connection with the carriers in the planes [15]. The vibrations of the plane oxygen atoms are expected to be sensitive to the carriers in the planes. The sudden increase of the width of the in-phase phonon, as well as the changes in its energy for x>6.3 are consistent with the introduction of carriers in the CuO_2 planes and a strong electron-phonon interaction for this mode. The coexistence of the high and low energy values for the in-phase phonon in the overdoped region and the sensitivity of the phonon to the buckling of the CuO_2 planes [1,14], indicates that there is an heterogeneous distribution of the two phases above optimum doping. It is unclear whether the carriers have the tendency to distribute inhomogeneously or other effects like strains induce the phase separation. Since lattice is strongly coupled to the carriers, the structural change will affect also the transition temperature.

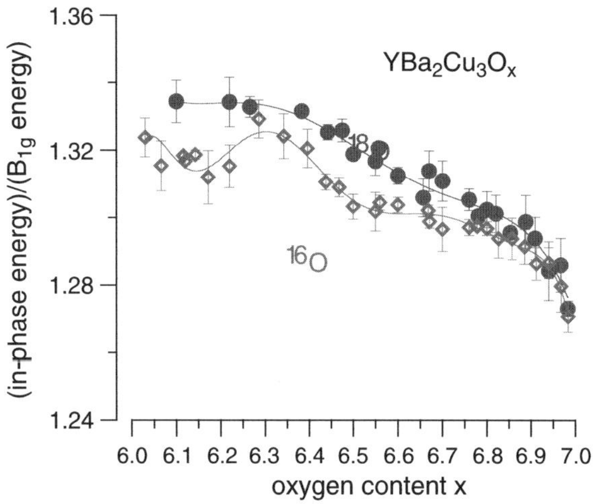

Figure III:2:3. The in-phase phonon energy normalized to the B_{1g}

The effect of the excess carriers induced by Ca substitution for Y up to the solubility limit of 20% has been investigated by Raman spectroscopy. In the xx scattering geometry of the Y-Ca123 compounds the phonon due to the out-of-phase vibrations of O_{pl} of B_{1g} symmetry, contains in addition to the well-known mode at 340 cm^{-1} another at ~322 cm^{-1} (Fig.4).

This mode becomes very strong in high Ca concentrations while a much weaker mode appears at ~296 cm^{-1} (Fig.4). The energy of the mode at 322 cm^{-1} remains independent of the amount of Ca and it has exactly the B_{1g}-like (Fig.4:Left). The other peak at ~340 cm^{-1} corresponds in energy to a pure undistorted Y123:x=opt compound [16,17].

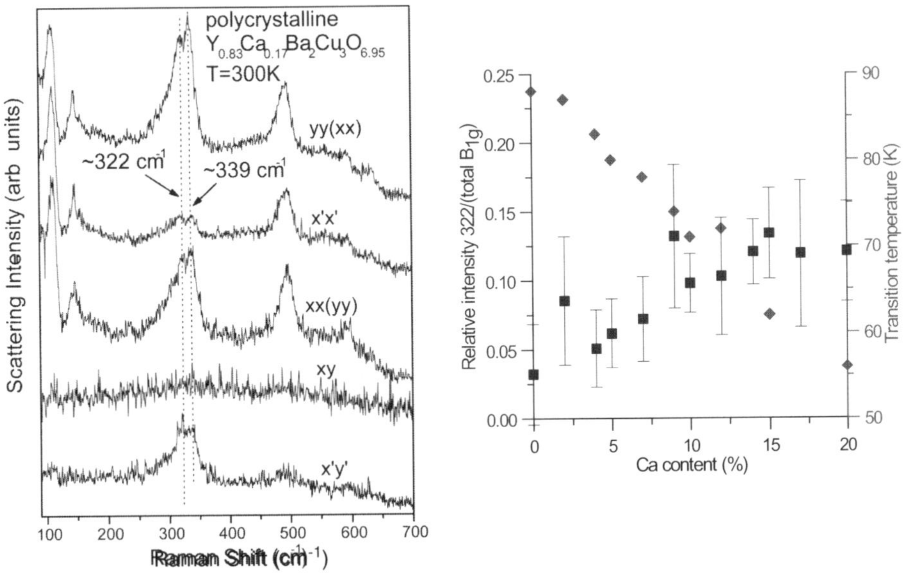

Figure III:2:4. Left: selection rules for the mode at 322 cm^{-1}, which prove its pure B_{1g} character for Ca concentration 17%. *Right*: The Ca dependence of the relative intensity of the 322 cm^{-1} mode to the total of the band, in connection with the transition temperature

In mixed compounds where Y has been substituted by a rare earth or a mixture of them ($R1_{1-y}R2_yBa_2Cu_3O_{7-\delta}$), there is a tendency for a separation into three phases; two of them correspond to the pure end compounds ($R1Ba_2Cu_3O_{7-\delta}$ and $R2Ba_2Cu_3O_{7-\delta}$) and another to an intermediate phase ($R1_{1-y}R2_yBa_2Cu_3O_{7-\delta}$), where y may not coincide with the nominal value [18-20]. The separation into phases is mainly driven by the internal strains from the difference in the ionic size of Y and the rare earths [18-20]. A similar effect should happen with the substitution of the larger Ca ion for the smaller Y. In Fig.4 (Left) the modes at ~340 cm^{-1} and ~322 cm^{-1} remain constant in energy independently of the amount of Ca [17]. The high-energy mode at ~340 cm^{-1} apparently corresponds to the pure Y123:x=opt phase and the other one at ~322 cm^{-1} should be the intermediate phase. The constant value of the intermediate mode energy would imply a constant ratio Y, Ca in this mixed phase not depending on the nominal value of Ca concentration. Based on the dependence of the B_{1g} phonon energy on the rare earth ion size [18-20], we can calculate that the energy of the ~322 cm^{-1} mode should correspond to a mixed $Y_{0.5}Ca_{0.5}Ba_2Cu_3O_{7-\delta}$ phase with approximately equal amounts of Ca and Y atoms. EXAFS measurements on the same samples have suggested that there are no Ca-Ca clusters and each Ca is surrounded by four Y atoms [1,21]. If the two results (local probe and collective) are combined, one concludes that there must be a separation into phases one of pure Y123:x=opt and another mixed one of equal amounts of Y and Ca. This separation should be independent of the amount of Ca. The absence of any sign of two chemical phases in XRD or neutron scattering measurements indicates that the intermediate phase should be in small clusters of a size of a few unit cells, enough to be observed by Raman spectroscopy, but indiscriminate by XRD or neutron scattering. In Fig.4 right it is seen that the amount of the intermediate phases correlates with the reduction of the T_c. This implies that the spatial heterogeneity modifies the local carrier concentration as well and the two phases have a different T_c. Such a phase separation makes doubtful the generally accepted picture presented in Ref.[22] about the Ca overdoping of Y123.

3.2 Hydrostatic pressure effects

The above-discussed cases of doping-induced phase separation indicate that the compounds examined are close to a structural instability at the optimal doping. The attempt to overdope either by oxygen or Ca induces a separation into phases with different T_c. In general, atomic substitutions may induce disorder and its effect cannot be estimated easily. An alternative way of altering the properties of the compound is by applying hydrostatic pressure. Although Raman measurements under hydrostatic pressure for such compounds with low scattering efficiency are rather difficult, a series of superconductors have been investigated [10-13]. In all cases studied (Y123:x~6.5, Y123:x~7, Y124, and Bi2212), a clear deviation from linear dependence on pressure has been observed for several modes and this deviation, though it was measured at room temperature (RT), correlates with changes in T_c with pressure.

Figures 5-6 show characteristic spectra at RT and the pressure dependence of the energy of the A_g-symmetry phonons for Y123:x~6.5 and x~7. The pressure dependence of the energy of the B_{1g}-like mode is almost linear for the yttrium-based compounds we have studied. As seen in Fig.6, the A_g phonons due to Ba, in-phase, and apex show a deviation from linearity for pressures ~2.0 GPa. The deviation from linearity for the mode due to Cu(2) is smaller and harder to discriminate. At about the same pressures the width of the phonons changes abruptly [12]. Especially the phonon of the in-phase vibrations of the O_{pl} appears as a double peak above the same pressures (Fig.5).

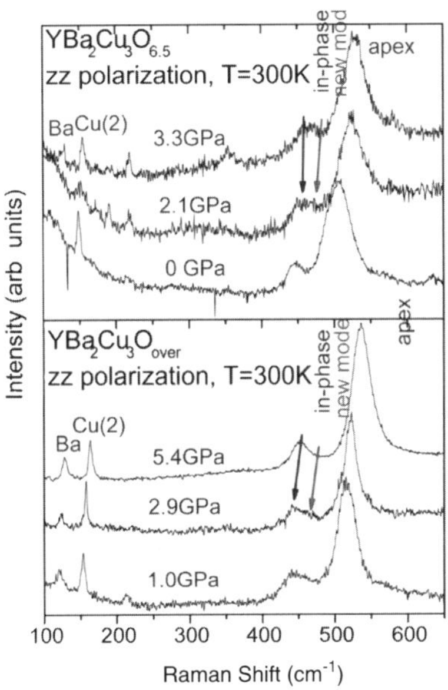

Figure III:2:5. Typical Raman spectra of the Y123:x~6.5 (top) and Y123:x~7 (bottom) compounds at various hydrostatic pressures in the xx/yy scattering geometry

The behaviour of the phonons can be related to pressure induced structural modifications. Unfortunately, for the Y123 compounds, the variation of the bond lengths as a function of pressure has been studied only for low pressures, up to 0.56 GPa [23] and there are no data available for an independent verification of the predictions of local structural modifications. The structural hydrostatic pressure measurements have found a similar pressure dependence of the a- and c-axes for the underdoped and overdoped Y123, while, for the b axis, the underdoped exhibits a larger compression (~20%) than the overdoped sample [23]. The Raman data for low pressures

(<2GPa) indicate the same rate of phonon energy increase with pressure for the two concentrations (Fig.6) in agreement with the XRD data.

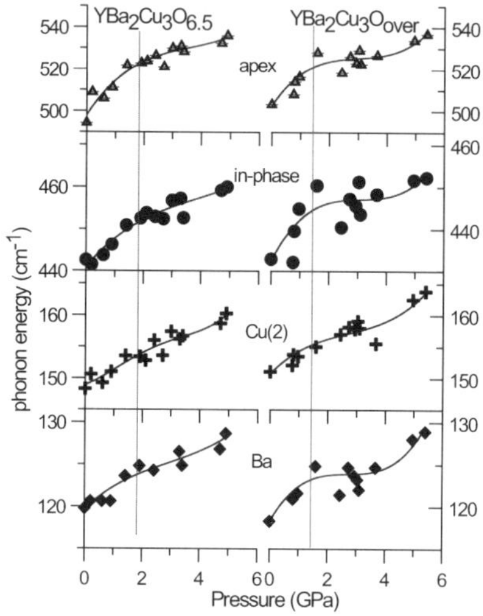

Figure III:2:6. The hydrostatic pressure dependence of the energy for the A_g Raman modes of the Y123x compounds. The lines are 3rd order polynomial best fits to the data

Ambient pressure studies have shown that the energy of the B_{1g}-like mode is independent of the oxygen concentration and the Cu_{pl}-O_{pl} bonds [4], but it is significantly affected by a rare earth substitution for Y showing a phase separation [19]. All available data indicate that the B_{1g} mode depends mainly on the Y-O_{pl} bonds [4,18-20]. Our Raman data show no anomaly for the pressure dependence of the B_{1g}-like mode energy for both Y123:x~6.5 and Y123:x~7 compounds [11-12], indicating that the Y-O(2,3) bonds do not undergo abnormal modifications with pressure in Y123 independently of the amount of doping.

According to lattice dynamical calculations, the in-phase phonon energy depends mainly on the plain Cu_{pl}-O_{pl} bonds [24] and this agrees with the ambient pressure studies of Y123x, where with increasing x the length of the Cu_{pl}-O_{pl} bonds increases almost linearly [1], while the in-phase mode energy decreases [4]. Therefore, the appearance of another mode with hydrostatic pressure is a sign that the Cu_{pl}-O_{pl} bonds of the planes undergo some modifications at a critical pressure, inducing a separation into phases.

The Y124 has double chains along the *b* axis giving stronger O_{ch}-Cu_{ch} bonds and diminishing the oxygen defects. Although the structure and T_c of Y124 is similar to Y123:x~7, the dT_c/dP is quite different from 5.5K/GPa for Y124 [25-27] to ~(1 to –2.5) K/GPa for Y123x [28]. Fig.7 presents typical Raman spectra of Y124 at high pressures and again considerable spectral modifications can be observed. For the pressure dependence of the energy of the modes, a deviation from linearity begins at ~3.8 GPa for several modes (Fig.8). At higher pressures (>5.5 GPa) the energy of the modes stops increasing (Fig.8). The modifications at ~3.8 GPa are accompanied by an increase of the width for the A_g-symmetry modes, while the $B_{2g,3g}$ phonons remain unaffected (Fig.7). In the case of the in-phase phonon the increase of the width is clearly due to the appearance of another mode at higher by ~20 cm^{-1} energy to the in-phase phonon (Fig.7).

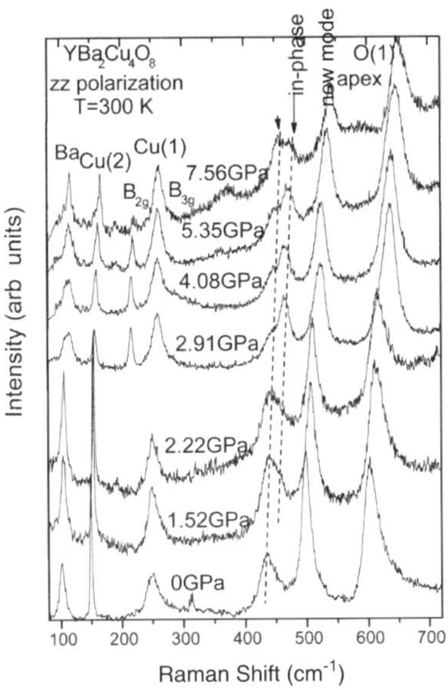

Figure III:2:7. Typical Raman spectra of the $YBa_2Cu_4O_8$ compound at selected hydrostatic pressures in the zz scattering geometry

The relative intensity of the in-phase to the new mode changes by rotating the crystal by a few degrees and keeping the same (zz) polarization for the incident and scattered light. It has been observed that the relative intensity of this new mode increases simultaneously with the intensity of the

B_{2g} phonon attributed to the vibrations of the O_{ch} atoms along the a-axis. It seems that the intensity variations and the appearance of the new mode are related with local distortions close to the CuO_2 planes and possibly a titling of the axis, which drives a phase separation in Y124 [11,12].

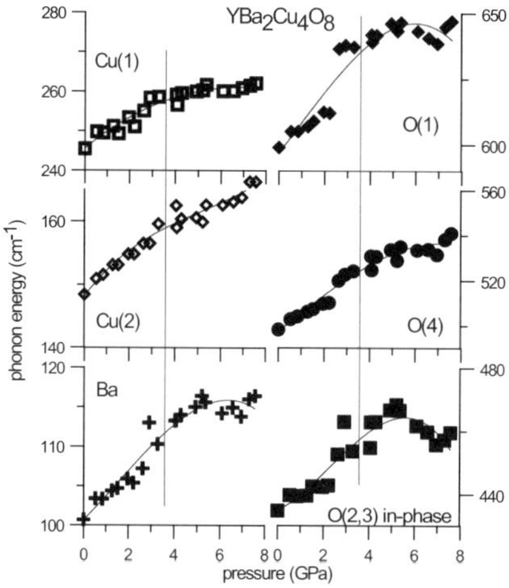

Figure III:2:8. Hydrostatic pressure dependence of the energy for the A_g-symmetry Raman modes of $YBa_2Cu_4O_8$. The lines are 3rd order polynomial best fits to the data

Structural data under hydrostatic pressure from XRD measurements of Y124 exist up to ~5 GPa [29]. They have shown, as expected, that the lattice parameters and all bond lengths decrease with increasing pressure, but the rate of change deviates from linearity above ~3.8 GPa [29]. At hydrostatic pressures ~5.5 GPa the dTc/dp changes rapidly and gradually the transition temperature saturates [26]. This behaviour is correlated with the changes observed in several phonons for the same pressure (Fig.8). Unfortunately, there are no structural data available above 5 GPa to investigate any anomaly on the interatomic bonds. One should remark once more that the presented Raman measurements are at RT and correlate with changes in T_c.

In Bi2212 there are many modes that are Raman active. Under hydrostatic pressures only the strong modes could be investigated [13]. At ambient pressure these are located at ~60 cm^{-1} (Bi), ~118 cm^{-1} (Sr), ~290 cm^{-1} (out of phase vibrations of O_{pl}, the B_{1g} mode), ~330 cm^{-1}, ~458 cm^{-1} and ~465 cm^{-1} (a doublet), and at ~630 cm^{-1}, (O_{Bi}). The assignment for certain modes is still controversial, but there is a general agreement for the low energy phonons at 60 cm^{-1} and 118 cm^{-1}, and the B_{1g} phonon [30-33]. The two strong modes at ~458 cm^{-1} and ~465 cm^{-1} appear with varying relative intensities with pressure. Both modes have shown a characteristic independence on pressure and they show an abnormal behaviour with temperature with a maximum energy around 150 K [13].

In Figure 9 we observe that the pressure dependence of the phonons due to Bi, Sr, the B_{1g}, and mainly the double mode at 458-465 cm^{-1} shows a strong modification at ~1.8 GPa.

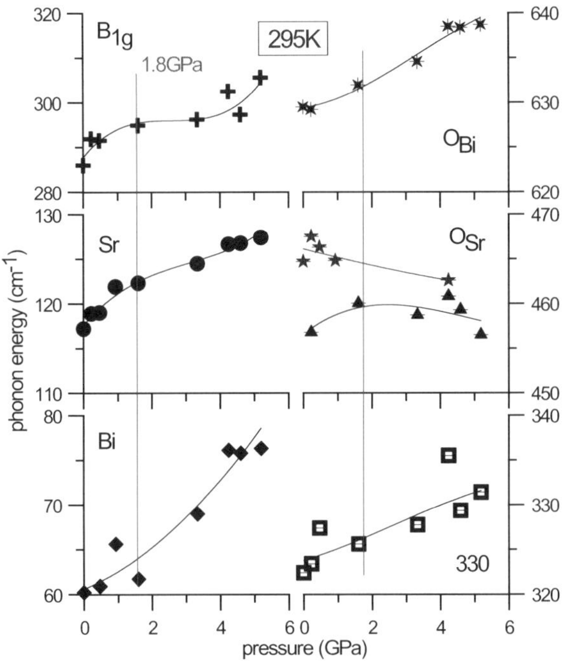

Figure III:2:9. The pressure dependence of the Raman active modes of Bi2212

At exactly the same hydrostatic pressure an anomaly has been measured for the transition temperature and it has been attributed to a deformation of the CuO_5 pyramids [31]. This structural modification affects several phonons but mainly the double peak at ~458-465 cm^{-1}. The splitting of this mode must again be related with a phase separation mechanism, similar to the ones discussed above for the other cuprates.

In Bi2212 there is an uncertainty about the assignment of the high-energy modes. The mode at 630 cm^{-1} appears also in the Bi2201 compound [34] and it cannot be associated with the in-phase phonon. Since it does not show any anomaly at 1.8 GPa pressure (Fig.9) when there is a structural distortion at the pyramids, it is improbable to be related with the apical oxygen. It is therefore reasonable to assign it to the oxygen atoms in the BiO planes [13]. The peak at ~330 cm^{-1} could be the in-phase mode since it shows a similar behaviour with the B_{1g} mode of the plane oxygen atoms [13]. This agrees with the association of the changes with an anomaly in the CuO_5 pyramids [31]. The doublet cannot be related with the O_{Bi} atom [30.31], since it should not be sensitive to the change in the pyramids. So, it has to be assigned to the vibrations either of the apical oxygen [13] or the in-phase phonon [33]. In either case, it is affected by the structural distortions of the CuO_5 pyramids and the double peak pointing to a phase separation scenario.

4. CONCLUSIONS

The presented Raman measurements, performed at RT, clearly show for certain phonons considerable deviations from the linear dependence on pressure. These modifications seem to originate from small structural changes induced by pressure. A similar small local structural modification that affects a certain phonon (in-phase) has been observed in Y123 at the boundary of the optimal to overdoped oxygen concentration, appearing as a first order phase transition [3,4]. EXAFS and neutron measurements have shown that it was due to a sudden change in the CuO_2 buckling and it was accompanied by a small change in T_c [1,3]. In another cuprate ($Bi_2Sr_2CaCu_2O_8$) a similar splitting of a mode was discovered indicating a separation into phases and correlated again with a change in the transition temperature [13]. In $La_{2-x}Sr_xCuO_4$ it was found that the T_c dependence on chemical doping follows spectroscopic changes observed also at room temperature [35]. All data obtained at ambient conditions indicate that the variation and the optimal values of T_c obtained by chemical doping or by varying an external factor like pressure, are affected by structural instabilities and phase separation.

REFERENCES

1. E. Kaldis, in Handbook of Physics and Chemistry of the Rare Earths, ed. K.A. Gschneidner jr and L. Eyring (North Holland, 2001).
2. J.B. Torrance, Y. Tokura, A.I. Nazzal, A. Benzike, T.C. Huang, S.S.P. Rarkin, Phys. Rev. Lett.61, 1127 (1988).
3. E. Kaldis, J. Röhler, E. Liarokapis, N. Poulakis, K. Conder, and P. W. Loeffen, Phys. Rev. Lett.79, 4894 (1997).
4. N. Poulakis, D. Palles, E. Liarokapis, K. Conder, E. Kaldis, and K.A. Müller, Phys. Rev. B53, R534 (1996).
5. D. Palles, D. Lampakis, N. Poulakis, K. Conder, E. Kaldis, and E. Liarokapis, Physica C282-287, 1051 (1997).
6. K. Conder, D. Zech, Ch. Krüger, E. Kaldis, A.W. Hewat, and E. Jilek, in Phase Separation in Cuprate Superconductors, eds. E. Sigmund and K.A. Müller, 210 (Springer, 1994).
7. J. Röhler, P. W. Loeffen, S. Mullender, K. Conder, and E. Kaldis in Workshop on High-Tc Superconductivity 1996: Ten Years after the Discovery, eds. E. Kaldis, E. Liarokapis, and K.A. Müller, (NATO ASI Series, Kluwer Academic Publishers, 1997).
8. J. Schilling and S. Klotz, in Physical Properties of High Temperature Superconductors, Vol. III, ed. D.M. Ginsberg (World Scientific, 1992).
9. J. Schilling, J. Phys. Chem. Solids, 59, 553 (1998).
10. E. Liarokapis, D. Lampakis, T. Nishizaki, and C. Panagopoulos, High Pressure Research 18, 109 (2000).
11. D. Lampakis, A. Tatsi, E. Liarokapis, and J. Karpinski, High Pressure Research 23, 123 (2003).
12. D. Lampakis, E. Liarokapis, J. Karpinski, C. Panagopoulos, and T. Nishizaki, to appear in J. of Superconductivity (2004).
13. E. Liarokapis, D. Lampakis, C. Panagopoulos, and T. Nishizaki, High Pressure Research 23, 111 (2003).
14. E. Liarokapis, D. Palles, K. Conder, E. Kaldis, J. of Raman Spectrosc. 32, 821 (2001).
15. T.P. Deveraux, A. Virosztek, and A. Zawadowski, Phys. Rev. B51, 505 (1995).
16. D. Palles, E. Liarokapis, Th. Leventouri, B.C. Chakoumakos, J. Phys.: Cond. Matter 10, 2515 (1998).
17. D. Palles, D. Lampakis, G. Boettger, E. Liarokapis, and E. Kaldis, Physica C341-348, 2149 (2000).
18. G. Bogachev, M. Abrashev, M. Iliev, N. Poulakis, E. Liarokapis, C. Mitros, A. Koufoudakis, and V. Psyharis, Phys. Rev. B49, 12151 (1994).
19. M. Calamiotou, A. Gantis, D. Palles, D. Lampakis, E. Liarokapis, A. Koufoudakis, Phys. Rev. B58, 15238 (1998).
20. Gantis, M. Calamiotou, D. Palles, D. Lampakis, E. Liarokapis, Phys. Rev. B68, 064502 (2003).

21. J. Röhler, C. Friedrich, T. Granzow, E. Kaldis, and G. Böttger, AIP Conf. Proc. 483, 320, HTS99, Miami January 1999, eds. S.E. Barnes et al., LANL cond-mat. 9902030.
22. L. Tallon, C. Bernhard, H. Shaked, R.L. Hifferman, J.D. Jorgensen, Phys. Rev. B51, 12911 (1995).
23. J.D. Jorgensen et al., Physica C171, 93 (1990).
24. C. Thomsen and M. Cardona, in Physical Properties of High Temperature Superconductors, ed. D.M. Ginsberg, World Scientific (1994).
25. B. Bucher, J. Karpinski, E. Kaldis, and P. Wachter, Physica, C, 157, 478 (1989).
26. E.N. van Eenige, R. Griessen, R.J. Wijngaardem, J. Karpinski, E. Kaldis, S. Rusiecki, and E. Jilek, Physica, C, 168, 482 (1990).
27. J.J. Scholtz, E.N. van Eenige, R.J. Wijngaarden, and R. Griessen, Phys. Rev., B, 45, 3077 (1992).
28. Driessen, R. Griessen, N. Koeman, E. Salomons, R. Brouwer, D. G. de Groot, K. Heeck, H. Hemmes, and J. Rector, Phys. Rev., B, 36, 5602 (1987); J. Karpinski, S. Rusiecki, E. Kaldis and E. Jilek, J. Less Com. Metals, 164 5, 3 (1990); S. Sadewasser, J.S. Schilling, A.P. Paulikas, and B.W. Veal, Phys. Rev., B, 61, 741 (2000).
29. R.J. Nelmes et al., Physica C172, 311 (1990).
30. M. Kakihana, M. Osada, M. Käll, L. Börjesson, H. Mazaki, H. Yasuoka, M. Yashima, and M. Yoshimura, Phys. Rev., B, 53, 11796 (1996).
31. M. Osada, M. Kakihana, M. Käll and L. Börjesson, Physica, C, 357-360, 142 (2001).
32. M. Cardona, C. Thomsen, R. Liu, H.G. von Schnering, M. Hartweg, Y.F. Yan, and Z.X. Zhao, Solid State Commun. 66, 1225 (1988).
33. M. Böckholt, G. Güntherodt, L.L. Leonyuk, V.V. Moshchalkov, Physica, C, 185-189, 1035 (1991).
34. M. Holiastou, N. Poulakis, D. Palles, E. Liarokapis, D. Niarchos, U. Frey, and H. Adrian, Physica C, 282-287, 583 (1997).
35. A. D. Lampakis, D. Palles, E. Liarokapis, C. Panagopoulos, J.R. Cooper, H. Ehrenberg, T. Hartmann, Phys. Rev. B62, 8811 (2000).

III.3

STRUCTURAL SYMMETRY, ELASTIC COMPATIBILITY, AND THE INTRINSIC HETEROGENEITY OF COMPLEX OXIDES

S. R. Shenoy[1], T. Lookman[2], A. Saxena[2], and A. R. Bishop[2]
[1]*International Centre for Theoretical Physics Trieste, Italy*
[2]*Los Alamos National Lab, Los Alamos, N.M., USA*

Abstract: Intrinsic heterogeneity, from nano- to meso- scales, of lattice/charge/spin variables, is common in complex oxides such as cuprates and manganites, that have strain-based ferroelastic transitions. We summarize our viewpoint that the central player is the power-law, anisotropic, strain-strain force that lies concealed in the apparently innocuous St. Venant compatibility constraint, namely the maintenance of lattice integrity, under strain texturing induced by charges and spins, that act as 'local stresses' and 'local transition temperatures'

Key words: Transition Metal Oxides, Elasticity, Compatibility, Cuprates, Manganites

1. INTRINSIC HETEROGENEITY

In complex electronic oxides such as cuprates and manganites, a variety of finescale probes have revealed an unsuspected nanoworld of correlated multiscale spatial patterning in lattice structure, charge density/mobility, and staggered/direct magnetization [1,2,3]. These sign-varying patterns or 'textures' have been referred to as stripes, islands, nanoscale phase-separations, etc. Interestingly, there are both direct and *cross*-responses to external fields; and local responses to global perturbations. Thus, in

A. Bianconi (ed.), Symmetry and Heterogeneity in High Temperature Superconductors, 133–145.

manganites not only does magnetic field change (colossally [1]) the resistance, but so does external stress [4]; and moreover a uniform field can induce nanoscale 'cloudlike' variations in the local conductance [5]. In cuprates (with an antiferromagnetic parent compound) a uniform magnetic field can orient mesoscale structural twins [6]. This argues for a multivariable cross-coupling: so external stress changes strain, that changes magnetization coupled to it; and similarly for other variables. Clearly an internal charge variation can act as an internal stress and affect the spins. The resultant self-consistent equilibrium state is not necessarily uniform. The *annealed*-variable inhomogeneities that may emerge from this internal coupling will here be termed 'intrinsic', distinguishing them from quenched-variable 'extrinsic' inhomogeneities also unavoidably present in the (usually nonstoichiometric) complex oxides. The fact that uniform fields can induce locally varying *multiscale* response argues for nonlinearities in at least one of these component variables, and so in Fourier space the resultant coupling of wave vectors interlinks the hierarchy of spatial (and temporal) scales.

Some of the extensive experimental and theoretical literature has been cited elsewhere [1,2]. Here, we focus on our own viewpoint regarding the underlying cause of these intrinsic inhomogeneities/heterogeneities [7-12]. The natural prejudice in favour of electronic/structural/magnetic states that are uniform at equilibrium, arises from the positive sign of gradient-squared terms: an initially random state flows energetically downhill by smoothing out spatial variations at all wavelengths. There are two exceptions to this: Nonlinear systems can have domains of competing ground states, with domain walls dividing the system–once formed, the nonuniform states are stable; and long-range forces competing with short-range couplings are well known to be able to induce nonuniform states. (Indeed, some models for stripes have invoked isotropic long-ranged Coulomb forces.) Our own Ginzburg-Landau model has *both* generic sources of nonuniformity. Apart from local couplings of charge and spin to strains, it has (structural-transition) nonlinearities in order-parameter strains; and purely local harmonic terms in other strains that, after enforcing a lattice integrity constraint [13-15], effectively become *anisotropic* long-range potentials. Our work explores the possibility that the constrained nonlinear system spontaneously forms multivariable inhomogeneous states, even without quenched disorder.

The key theoretical idea is to work in the *strain* representation [7-12,15] and impose the constraint of lattice integrity at all scales, through the St.

Venant compatibility (differential) equations for the strain-tensor components [14]. Elimination of some harmonic contributions in the free energy to satisfy the compatibility constraint leads to power-law anisotropic (PLA) potentials between the remaining order-parameter strain components, that vary directionally in sign [7,16]. Hence a local strain can receive conflicting ('ferro/antiferro') instructions from surrounding and even far-off strains, in a kind of elastic 'frustration' facilitating a structural patterning, and generating configurational glass-like multiple minima (a "landscape") in the space of all such possible states. Other variables coupled to lattice strains will also show sign-varying patterns or 'textures', through induced St. Venant PLA interactions. Thus in our scenario, strain modulations of an anisotropic, nonlinear, and non-ruptured lattice occur in its response to multiple local nanostresses exerted by the charges and spins; and these in turn, have density-wave modulations induced by the lattice strain textures [10,11]. The self-consistent multi-variable equilibrium states in doped ferroelastics such as the complex oxides can be intrinsically heterogeneous.

2. LATTICE HAMILTONIANS: SPRINGS AND STRAINS

As noted elsewhere [11], it is useful to classify *levels* of description through inverse wave vector scales $\lambda \equiv k^{-1}$, of solids with system size L_o, Bravais lattice until-cell a_o, and atomic basis scale l_o. Level 0 is the $\lambda \leq L_o$ scale of engineering mechanics; Level 1 is the (large) range of lattice strain variations $L_o > \lambda > a_o$; Level 2 is the intra-cell bond-stretching and bending or 'microstrain' range $a_o > \lambda > l_o$; and Level 3 is the electronic orbital scale $l_o > \lambda$. Consider for simplicity a 2D square unit-cell lattice $a_o \equiv 1$, with a monatomic basis at positions $\{\vec{R}_i\}$. Then the general interatomic potential energy is $V = (1/2)\sum_{i,j} V(\vec{R}_i - \vec{R}_j) = (1/2)\sum_{i,j} V(\vec{r}_i - \vec{r}_j + \vec{u}(\vec{r}_i) - \vec{u}(\vec{r}_j))$ where ith atomic positions displaced from reference lattice sites $\{\vec{r}_i\}$ are $\vec{R}_i = \vec{r}_i + u(\vec{r}_i)$. Taylor expanding in the displacement differences with $\vec{\Delta}$ the discrete difference operator $\Delta_\mu f(\vec{r}) \equiv f(\vec{r} + \hat{\mu}) - f(\vec{r})$ and $\hat{\mu} = \hat{x}, \hat{y}$, and with nearest-neighbour couplings only retained,

$$V \simeq (1/2)\sum_i \sum_{\substack{\mu\mu' \\ \nu\nu'}} C_{\mu\nu,\mu'\nu'} \Delta_\mu u_\nu \Delta_{\mu'} u_{\nu'} + \dots = (1/2)\sum_i \sum_{\substack{\mu\mu' \\ \nu\nu'}} C_{\mu\nu,\mu'\nu'} E_{\mu\nu} E_{\mu'\nu'} + \dots \quad (1)$$

where with symmetric 'spring constant' coefficients $C_{\mu\nu,\mu'\nu'}$ it is convenient to define a symmetric *strain tensor* that in coordinate and Fourier space is

$$E_{\mu\nu}(\vec{r}) \equiv (1/2)[\Delta_\mu u_\nu(\vec{r}) + \Delta_\nu u_\mu(\vec{r})] \, ; \, E_{\mu\nu}(k) \equiv (i/2)[K_\mu u_\nu(\vec{k}) + K_\nu u_\mu(\vec{k})] \, , \quad (2)$$

with $iK_\mu(\vec{k})$ the Fourier transform of Δ_μ, and $K_\mu \simeq k_\mu$ at long wavelengths. The symmetric strain tensor $\underline{\underline{E}}(\vec{r})$ with $d(d+1)/2$ components describes deformations of, as well as variations between, the unit cell(s); and can be assigned to sites on the dual lattice e.g. at centroids of a square unit cell. The strain components transform as a second rank tensor: if (x, y) are coordinate variables, then clearly $E_{xx} \sim xx$, $E_{xy} \sim xy$ etc., under the unit-cell symmetry group. The physical strains are linear combinations of these.

The 1D case is of course trivial. In 2D there are $d(d+1)/2 = 3$ physical strains; compressional $e_1 = (E_{xx} + E_{yy})/\sqrt{2}$ (e.g. square to larger/smaller square); $e_2 = (E_{xx} - E_{yy})/\sqrt{2}$ (e.g. square to rectangle, along x or y directions); and shear $e_3 \equiv E_{xy} \equiv E_{yx}$ (e.g. square to equal-side parallelogram). The strains are as defined even for other transitions, and combinations of e_2, e_3 describe deformations of an equilateral to isosceles triangle [8]. In 3D (cubic lattice) there are $d(d+1)/2 = 6$ strains [9,12,14]: one compressional $e_1 = (E_{xx} + E_{yy} + E_{zz})/\sqrt{3}$; two independent deviatoric strains $e_2 = (E_{xx} - E_{yy})/\sqrt{2}$ and $e_3 = (E_{xx} + E_{yy} - 2E_{zz})/\sqrt{6}$; and three shear strains $e_4 = E_{yz}$, $e_5 = E_{xz}$, $e_6 = E_{xy}$. (Here, the numerical prefactors are conventional [17].) In the continuum limit, the discrete differences become derivatives yielding engineering strains [14], but as defined, $\underline{\underline{E}}(\vec{r})$ applies throughout Level 1. The usefulness of the unit-cell strain concept becomes clear when modeling ferroelastic structural transitions.

3. STRUCTURAL SYMMETRY AND FREE ENERGY

The network of directional bonds linking atomic bases can soften with temperature, driving structural phase transitions. In complex oxides, ferroelastic transitions can occur, with strain (Level 1) as the order parameter, and unit cells deforming in a smooth displacive manner, with the low temperature strained-cell symmetry group being a subgroup of the high temperature (zero-strain) unit-cell symmetry group [12]. The microscopic partition function will involve a trace over all the electronic and nuclear degrees of freedom (Level 3) that form the directional bonds of the atomic basis (Level 2), consistent with the strained lattice (Level 1). The resulting free energy functional F is a scalar function of the local strain tensor, with a sum over unit-cell contributions, i.e. $F = \sum_r f(\underline{\underline{E}}(\vec{r}), \vec{\Delta}\underline{\underline{E}}(\vec{r}))$. The free energy density f can be written as a series in increasing-order polynomials in the strain-tensor components, that are invariant under the *discrete* (high temperature) unit-cell symmetry group. For the square unit cell, and with $x = R\cos\theta$, $y = R\sin\theta$, examples of invariants are $e_1^2 \sim (xx+yy)^2 \sim R^2$; $e_2^2 \sim (xx-yy)^2 \sim R^2\cos^2 2\theta$; $e_3^2 \sim (xy)^2 \sim R^2\sin^2 2\theta$; and a gradient term $\vec{\Delta}e_2 \cdot \vec{\Delta}e_2 \sim R^2\cos^2 2\theta$, all unchanged under e.g. 90° rotations, $\theta \to \theta + \pi/2$. For a triangular unit cell, invariants include the cubic, $e_2^3 - 3e_2e_3^2 \sim R^3\cos 3\theta$, unchanged under 120° rotation, $\theta \to \theta + 2\pi/3$.

A first-order square to rectangle transition, with deviatoric strain order parameter $\varepsilon \equiv e_2$ positive (negative) for rectangle along the x (y) direction. The invariant polynomials up to sixth power in the order parameter can be written in scaled form [7,8] as:

$$f(e_1, \varepsilon, e_3) = \xi_o^2(\vec{\Delta}\varepsilon)^2 + f_{non}(e_1, e_2) + f_o(\varepsilon). \quad (3)$$

Here the triple-well Landau term $f_o = (\tau - 1)\varepsilon^2 + \varepsilon^2(\varepsilon^2 - 1)^2$ has minima at $\varepsilon = \pm 1$, that are degenerate with the square unit-cell $\varepsilon = 0$ minimum at scaled temperature $\tau(T) = 1$. The non-order parameter term

$$f_{non} = (1/2)A_1 e_1^2 + (1/2)A_3 e_3^2 \quad (4)$$

is harmonic in the compressional and shear strains. Notice that in the displacement representation, every term in Equation (3) and (4) is an anisotropic gradient of $\vec{u}$: for example, $e_3^2 \sim (\Delta_x u_y + \Delta_y u_x)^2$ is quadratic, and $\varepsilon^6 \sim (\Delta_x u_x - \Delta_y u_y)^6$ is a sixth-power, displacement gradient term. Monte Carlo simulations [16] or free energy minimizations [18], in $\vec{u}$, yield 'twin' domain walls oriented along a diagonal.

Now comes our crucial point. It is tempting to think of the strains as the physical variables in a 'strain representation' for free energy minima: then since the minima of the harmonic Equations (4) are $e_1 = e_3 = 0$, one could further minimize $f(0,\varepsilon,0)$. However, this is incorrect as strains are all components of a single, symmetric strain tensor; that moreover has too many components: $d(d+1)/2 > d$, the number of displacement degrees of freedom. There must be $N_c = d(d-1)/2$ independent constraints, to obtain the correct $d(d+1)/2 - N_c = d$ number of degrees of freedom. In fact, these required constraints for the strain representation to be consistent, do exist: they are the St. Venant compatibility conditions [13,14,15], expressing the single-valuedness of the underlying displacement field, with no defects.

4. ELASTIC COMPATIBILITY: LATTICE INTEGRITY EQUATIONS

The St. Venant compatibility equations [13,14,15] follow immediately from the strain tensor definition of Equation (2), and are

$$\vec{\Delta} \times \{\vec{\Delta} \times \underline{\underline{E}}(\vec{r})\}^T = 0\,; \vec{K} \times \underline{\underline{E}}(\vec{k}) \times \vec{K} = 0 \tag{5}$$

where T means 'transpose'. As Baus and Lovett have noted [15], the displacement representation is similar to doing electromagnetism in terms of the vector potential $\vec{A}$, with Equation (5) then being just a vector identity analogous to $\vec{\Delta} \cdot (\vec{\Delta} \times \vec{A}) = 0$. However, once a physical field variable $\vec{B} = \vec{\Delta} \times \vec{A}$ is defined, it is illuminating to think of $\vec{\Delta} \cdot \vec{B} = 0$ as an independent field equation in a magnetic field representation. Similarly [7,8,9,11], the St. Venant constraint Equation (5) can be elevated to a field equation for the tensorial variables, in a *strain* representation.

In 2D, there are $N_c = 1$ compatibility equations [7,16]:

$$\vec{\Delta}^2 e_1 - (\Delta_x^2 - \Delta_y^2)\varepsilon - \sqrt{8}\Delta_x\Delta_y e_3 = 0\,; Q_1 e_1(\vec{k}) + Q_2\varepsilon(\vec{k}) + Q_3 e_3(\vec{k}) = 0 \qquad (6)$$

where $Q_{1,2,3}(\vec{k})$, the Fourier transforms of the difference operators, are more simply stated in their long wavelength limits, $Q_1(\vec{k}) \simeq -k^2/\sqrt{2}$; $Q_2(\vec{k}) \simeq k_x^2 - k_y^2/\sqrt{2}$; $Q_3 \simeq 2k_x k_y$.

If we substitute $e_1(\vec{k}) = -[Q_2\varepsilon(\vec{k}) + Q_3 e_3(\vec{k})]/Q_1$ into $f_{non}(e_1, e_3)$ of Equations (3), (4), then compatibility is satisfied for *any* $\varepsilon(\vec{k})$, $e_3(\vec{k})$. Minimizing with respect to $e_3(\vec{k})$ [19] yields non-OP strains in terms of the OP strain $e_{1,3}(\vec{k}) = B_{1,3}(\vec{k})\varepsilon(\vec{k})$, and so the free energy density of Equation (3) is entirely in terms of the OP strain. The harmonic non-OP cost becomes a *nonlocal* PLA compatibility potential term, $f_{non}(e_1, e_3) \equiv f_{compat}(\varepsilon)$ so Equation (3) becomes

$$f(\varepsilon) = \xi_o^2(\vec{\Delta}\varepsilon)^2 + f_{compat}(\varepsilon) + f_o(\varepsilon), \qquad (7a)$$

where in Fourier space

$$f_{compat} = (1/2)[A_1 |B_1(\vec{k})|^2 + A_3 |B_3(\vec{k})|^2]|\varepsilon(\vec{k})|^2 \equiv (1/2)A_1 U(\vec{k})|\varepsilon(\vec{k})|^2 \qquad (7b)$$

and in coordinate space $f_{compat} = (1/2)A_1 U(\vec{r} - \vec{r}')\varepsilon(r)\varepsilon(r')$. The compatibility kernel for the 2D square-rectangle transition is (with $D(\vec{k}) \equiv Q_3^2 + (A_3/A_1)Q_1^2$):

$$U(\vec{k}) = \frac{(A_3/A_1)Q_2^2}{D}; e_1(\vec{k}) = -\frac{(A_3/A_1)Q_2 Q_1 \varepsilon(\vec{k})}{D}; e_3(\vec{k}) = -\frac{Q_2 Q_3 \varepsilon(\vec{k})}{D}. \qquad (8)$$

The free energy minima can be found by a relaxational OP dynamics [8] $\partial\varepsilon/\partial t = -\partial F/\partial\varepsilon$ with initial conditions $\varepsilon(\vec{r}, t=0)$ random, with zero mean. In this (OP) strain representation, the $\varepsilon(\vec{r})$ variables are 'effective scalars' at each site, with anisotropy of the fourfold unit-cell symmetry carried entirely by the compatibility potential $U(\vec{r}-\vec{r}\,') \sim \cos 4(\theta-\theta')/|\vec{r}-\vec{r}\,'|^2$ where $\hat{r}\cdot\hat{r}'=\cos(\theta-\theta')$. The power law decay $\sim 1/R^d$, is faster than Coulombic (or critical correlation-like) fall-off $\sim 1/R^{d-2}$, and arises generically from the Fourier transform of kernels as in Equation (8) that for long wavelengths are independent of $|\vec{k}|$ since they depend only on *ratios* $Q_{2,3}/Q_1$: as $U(\vec{k})\simeq U(\hat{k})$, one has $\int d^d k e^{i\vec{k}\cdot\vec{R}} U(\vec{k}) \sim 1/R^d$. Since the angular average of $U(\vec{R})$ is zero, the influence of arbitrary strain variations could fall off even faster.

Each ferroelastic transition has its own characteristic compatibility kernel, that can be evaluated in 2D [7,8,12,16] and 3D [9]. The sign variation with direction of $U(\vec{r}-\vec{r}\,')$ implies local strain has 'ferro/antiferro' (elastic) frustration, that tends to favour spatial strain texturing, or patterns of domain walls. Since $\sum_r f_{compat} > 0$ from its origin in Equation (4), for $A_1 \gg 1$ the St. Venant forces exert a strong orientational effect on domain walls, to reduce the local non-OP strains as much as possible. From Equation (8), $U(k) \sim (k_x^2 - k_y^2)^2$ so $\hat{k}_x \pm \hat{k}_y$ or $\pm\pi/4$ orientation domain walls are favoured for the square-rectangle transition with consequent expulsion of non-OP strains as $e_{1,3}(\vec{k}) \sim k_x^2 - k_y^2 \sim 0$ in an 'elastic Meissner effect' [7]. (For the triangle-rectangle transitions, the kernel favours three $2\pi/3$ orientation domain walls meeting at a point, around which the non-OP compressional strain is concentrated [8], as required for lattice integrity.) kernels in 3D induce complex helical texturing [9].

Domain walls N_{dom} of thickness ξ_o, across system size L_o, will reduce the bulk energy $-D_o L_o^2$ (where D_o is an elastic energy density) in a strip $\xi_o L_o$ and have a 'gradient squared' energy cost, so the textured-state energy is $-D_o[L_o^2 - N_{dom}\xi_o L_o] + D_o N_{dom}\xi_o^2(1/\xi_o^2)\xi_o L_o$. This is a fractional energy difference of only $\delta \equiv 2N_{dom}(\xi_o/L_o)$ above the uniform $N_{dom}=0$ state [20]. (For equal-width twins of separation $W \sim L_o^{1/2}$ [7], and $N_{dom} = L_o/W$ textured state, the difference $\delta \sim 1/L_o^{1/2}$.) In general, the uniform 'true' ground state is only one out of many asymptotically degenerate competing textured states, and the latter will be favoured, on quenching from an arbitrary initial state, especially as ferroelastic dynamics favours an early lock-in of large wave-vector textures [8]. For doped ferroelastics, charges and spins acting as local internal stresses and temperatures can make structural heterogeneity the preferred state.

5. COUPLING TO CHARGES AND SPINS

External stress, locally applied, can have *nonlocal* static effects in ferroelastics (see Fig. 4 of Ref. [7]). Dynamical evolution of strains under local external stress can show striking time-dependent patterns such as 'elastic photocopying' of the applied deformations, in an expanding texture (see Fig.5 of Ref. [8]). Since charges and spins can couple linearly to strain, they are like *internal* (unit-cell) local stresses, and one might expect extended strain response in *all* (compatibility-linked) strain-tensor components. Quadratic coupling is like a local transition temperature. The model we consider is a (scalar) free energy density term

$$f_{coupling} = A_{n1} n e_1 + A_{m1} m^2 e_1 + A_{n\varepsilon} n \varepsilon^2 + A_{nm} n m^2 , \tag{9}$$

where $n(\vec{r})$ is a sum of doped-charge number densities with an exponential profile $\exp(-|\vec{r}-\vec{r}_i|)$, with unit-cell extent around site $\vec{r}_i$. The magnetization variable $m(\vec{r})$ is the staggered magnetization for cuprates, and the (core-level) ferromagnetic moment for manganites, with

$$f_{mag} = d_o[\xi_m (\vec{\Delta} m)^2 + (T - T_{cm}) m^2 + (1/2) m^2] , \tag{10}$$

where d_o is a magnetic/elastic energy ratio; T_{cm} is a nonzero scaled Ne'el temperature for 'cuprates'; and $T_{cm} = 0$ for 'manganites' (purely paramagnetic parent compound). The constants $A_{n1}, A_{m1}, A_{n\varepsilon}, A_{nm}$ are chosen to mimic cuprates (manganites), e.g. $A_{nm} > 0$, ($A_{nm} < 0$), representing a suppressing (inducing) of magnetization by the hole-doped mobile charges. In a relaxational dynamics ($\partial\varepsilon / \partial t = -\partial F / \partial \varepsilon$, $\partial m / \partial t = -\partial F / \partial m$) and $f = f_{grad} + f_o + f_{compat} + f_{coupling} + f_{mag}$, with fixed charges randomly placed, we obtain $\varepsilon(\vec{r}), e_1(\vec{r}), m(\vec{r})$ textures [10,11].

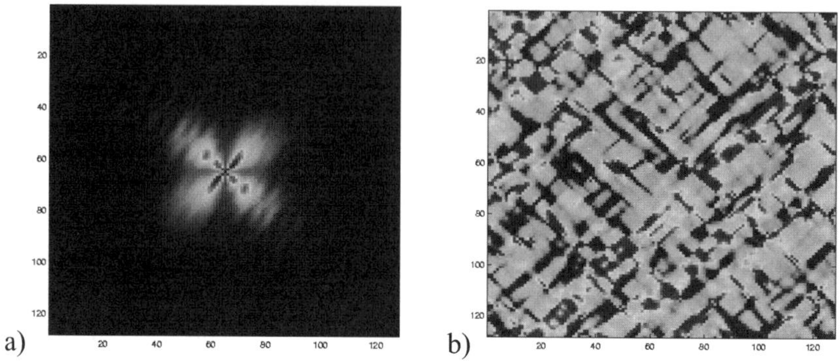

Figure III:3:1. a) Single 'cuprate' pemton greyscale plot of deviatoric strain $|\varepsilon(\vec{k})|^2$ in Fourier space. b) Multipemton greyscale deviatoric strain $\varepsilon(\vec{r})$ plot in coordinate space for hole-doped 'mobile' charges fraction x = 0.1

The textured polaron in $\varepsilon(\vec{r})$ that results, extends over tens of lattice spacings, with a fourfold symmetry from the PLA compatibility potential also reflected in Fourier space, Fig. 1a), cuprate case. Similar quadrupolar patterns occur in other coupled variables, and we term the excitation as a '*p*olaronic *e*lasto-*m*agnetic *t*exture' or '*pemt*on'. Fig 1b) shows diagonal stripe-like [3] textures for random doping fraction $x = 0.1$: the multipemtons are not rigid, but mutually deform each other. To get a complete picture both coordinate and Fourier space plots are necessary, and the logscale Fourier plots ($|\varepsilon(\vec{k})|^2, |m(\vec{k})|^2$) in Ref [11] reveal $\vec{k} \to -\vec{k}$ inversion symmetry even in fine details. The fourfold butterfly-like plots in experimental diffuse X-ray scattering in manganits and cuprates [21] (and their cross-response to a magnetic field) could possibly reflect such pemton signatures. Our model's multipemton textures show both direct and cross-responses, and at multiple scales: a long wavelength external magnetic field (stress) produces short-wavelength 'cloud-like' [5] gradations in strain (magnetization) [22]. This is similar to experiment [4,5,6] and occurs because of nonlinearity, with $\varepsilon^6(r) \to \prod_{s=1}^{6} \varepsilon_{qs}$, $m^4(r) \to \prod_{s=1}^{4} m_{qs}$ in Fourier space coupling wavevectors for different scales.

Charge and strain couple to magnetization quadratically ($\sim m^2$) and act as local variations of the transition temperature: from Equation (9) and (10), $T_{cm}'(r) = T_{cm} - (A_{m1}e_1 + A_{nm}m)/d_o$. Thus at a given T, some regions will be below, $T - T_{cm}'(r) < 0$ (above, $T - T_{cm}'(r) > 0$) transition, with $m^2(\vec{r}) \neq 0$ ($m^2(\vec{r}) = 0$), as seen in the large-fluctuation $m(\vec{r})$ (flat $m(\vec{r}) = 0$) regions in Fig. 2 of Ref. [10]. For manganites ($T_{cm} = 0$, paramagnetic parent) at low doping the spatial average $< m >= 0$ until a threshold doping x_c is reached when symmetry-breaking $< m > \neq 0$ occurs in these nonzero $m^2(\vec{r})$ regions. For warming at fixed $x \equiv < n >$, the flat $m(\vec{r}) = 0$ regions expand and $< m >$ decreases. Thus there is doping and temperature-dependent *annealed and correlated percolation*: this differs from quenched-disorder scenarios [1]. For cuprates ($T_{cm} \neq 0$), the model suggests the Ne'el temperature transition could have a percolative character.

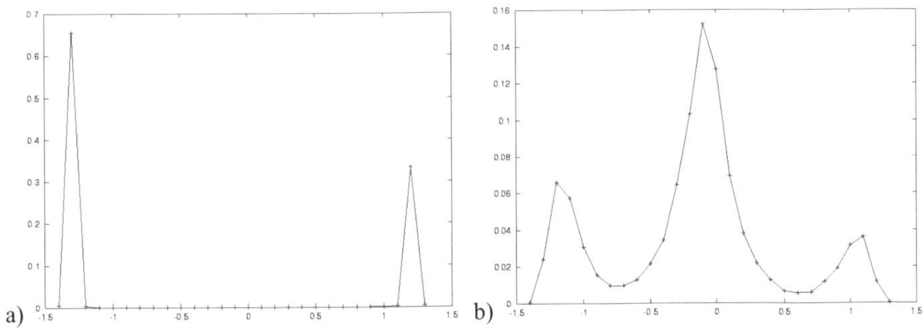

Figure III:3:2. a) For the 'twinned' parent compound, x=0, probability of finding a deviatoric strain $\varepsilon(\vec{r})$ versus the strain. b) Similar plot for x=0.1 multipemton strain

Fig. 2(a) and 2(b) show further results not presented in Ref. [10]: the probabilities of finding (positive/negative) strains of a given value show a double-hump behaviour. From the charge-strain couplings one might expect single-sign strains. However, *both* signs of deviatoric strain (and *both* compression and dilatation) occur, in energy-lowering 'adaptive elastic screening' [11]. For cuprates and manganites, bimodal distributions of interatomic distances are experimentally seen [3,23]. Although our model here [10] has fixed charges, an extension to allow hopping would result in charge ordering through $U(\vec{R})$ mediated charge-charge PLA forces.

Finally we comment that adding isotropic Coulomb forces should not destroy the essential features of the anisotropic intrinsic heterogeneity here induced by St. Venant forces alone. Conversely, any consistent calculation of dissipative response (e.g. resistivity) from Coulombic plus phononic scattering in the displacement representation will be unchanged by working in the strain representation, as the PLA elastic forces are already implicit in the generic assumption of lattice integrity, for both representations. Such compatibility contributions can be large for materials with widely differing strain-component energy scales, e.g. with dimensionless elastic constant $A_1 \gg 1$ [8]. This anisotropic (Level 1) parameter regime could be accessible in complex compounds with anisotropic (Level 2) atomic bases, e.g. in transition-metal oxides with perovskite octahedra' comprised of directionally bonded ions and deformable/polarizable oxygens. It is the interplay between strain nonlinearity, elastic compatibility, and multivariable couplings that drives the multiscale intrinsic heterogeneity of cuprates, manganites, and other complex adaptive materials.

REFERENCES

1. E. Dagotto, (Ed.) Nanoscale Phase Separation and Colossal Magnetoresistance, Springer (2003).
2. A.R. Bishop, S.R. Shenoy and S. Sridhar (Eds.) Intrinsic Multiscale Structure and Dynamics in Complex Electronic Oxides, World Scientific (2003).
3. A. Bianconi, N.L. Saini, A. Lanzara, M. Missori, T. Rosetti, H. Oyanagi, H. Yamguchi, K. Oka and T. Ito, Phys. Rev. Lett. 76, 3412 (1996); E.S. Bozin, G.H. Kwei, H. Takagi and S.J.L. Billinge, Phys. Rev. Lett. 84, 5856 (2000); K.M. Lang, V. Madhavan, J.E. Hoffman, E.W. Hudson, H. Eisaki, S. Uchida and J.C. Davis, Nature 415, 412 (2002).
4. Y. Huang, I.M. Palstra, S.W. Cheong and B. Batlogg, Phys. Rev. B 52, 15046 (1995)
5. M. Faeth, S. Friesen, A.A. Menovsky, Y. Tomloka, J. Aarts and J.A. Mydosh, Science 285, 1540 (1999).
6. Y. Ando in Ref. [2].
7. S.R. Shenoy, T. Lookman, A. Saxena and A. R. Bishop, Phys. Rev. B 60, R12537 (1999).
8. T. Lookman, S. R. Shenoy, K.O. Rasmussen , A. Saxena and A.R. Bishop, Phys. Rev. B, 67, 024114 (2003); T. Lookman, S. R. Shenoy, K.O. Rasmussen, A. Saxena and A.R. Bishop, J. de Physique IV, ICOMAT-02 Proc. (2003).
9. K.O. Rasmussen, T. Lookman, A. Saxena, A.R. Bishop, R. C. Albers and S.R. Shenoy, Phys. Rev. Lett. 87, 055704 (2001).
10. A.R. Bishop, T. Lookman, A. Saxena and S.R. Shenoy, Europhysics Lett. 63, 289 (2003).

11. S.R. Shenoy, T. Lookman, A. Saxena and A.R. Bishop, in Ref. [2].
12. A. Saxena, T. Lookman, A.R. Bishop and S.R. Shenoy, in Ref. [2]; D.M. Hatch, T. Lookman, A. Saxena and S.R. Shenoy, Phys. Rev. B 68, 104105 (2003).
13. A.J.C. Barre de Saint-Venant in C.L.M.H. Navier, Resume' des Lecons sur l'Application de la Mechanique (Dunod, Paris, 1864).
14. S.F. Borg, Fundamentals of Engineering Elasticity, Second edn. 1990, World Scientific.
15. M. Baus and R. Lovett, Phys. Rev. Lett. 65, 1781 (1990); 67, 406 (1991); Phys. Rev. A 44, 1211 (1991).
16. S. Kartha,, J.A. Krumhansl, J.P. Sethna and L.K. Wickham, Phys. Rev. B 52, 803 (1995).
17. The 'normalization' factors are chosen so that for e_1, e_2 in 2D $(E_{xx}{}^2 + E_{yy}{}^2)/2 = 1$; or for e_3 in 3D, $(4E_{xx}{}^2 + E_{yy}{}^2 + E_{zz}{}^2)/6 = 1$, at $|E_{xx}| = |E_{yy}| = |E_{zz}| = 1$.
18. A.E. Jacobs, Phys. Rev. B 52, 6327, (1995).
19. This direct method yields the same results as the Lagrange multiplier method for finding the kernel introduced in Ref. [16] for the square-rectangle case, and used for kernels in 2D and 3D in Refs. [7-12].
20. In Fig. 8 of Ref. [8], different nearly-degenerate textured states evolved under relaxational dynamics are shown. In scores of such relaxations, we have found a uniform state only once, for a particular initial-state random seed.
21. L. Vasiliu-Doloc, S. Rosenkranz, R. Osborn, S.K. Sinha, J.W. Lynn, J. Mesot, O.H. Seeck, G. Preost, A.J. Fedro and J.F. Mitchell, Phys. Rev. Lett. 83, 4393 (1999); S. Shimomura, N. Wakabayashi, H. Kuwahara and Y. Tokura, Phys. Rev. Lett. 83, 4389 (1999); Z. Islam, X. Liu, S.K. Sinha, J.C. Lang, S.C. Moss, D. Haskel, G. Srajer, P. Wochner,D.R. Lee, D.R. Haeffner and U. Welp, Phys. Lett. 93, 157008 (2004).
22. A.R. Bishop, T. Lookman, A. Saxena and S.R. Shenoy, cond-mat/0304198.
23. Ch. Renner, G. Aeppli, B-G. Kim, Y-A. Soh, and S.W.Cheong, Nature 416, 518 (2002); S.J.L. Billinge in Ref [2].

III.4

A CASE OF COMPLEX MATTER: COEXISTENCE OF MULTIPLE PHASE SEPARATIONS IN CUPRATES

G. Campi and A. Bianconi
Department of Physics, University of Rome "La Sapienza", P.le Aldo Moro 2, 00185 Rome, Italy

Abstract: A modified Van der Waals scheme for cuprates to give the co-existence of multiple phase separations in cuprates is presented. The model includes the tendency of charge carriers to form anisotropic directional bonds at preferential volumes for the formation of different "liquid phases". We obtain the variation of the pseudogap temperature $T^*(\delta)$ (for phase separation between pseudogap matter and normal matter) with hole density (δ) in agreement with experiments. We discuss the thermodynamic parameters that control the variation of the phase diagram of different cuprate families.

Key words: Multiple Phase Separation, Spinodal Lines, Cuprates

PACS numbers 64.75.+g, 81.30.Mh, 74.72.-h

1. INTRODUCTION

The problem of phase separation in cuprates superconductors has been longely debated [1-8]. Recently, several experiments show the formation of electronic crystals at critical densities [9, 10]. These results provide a strong experimental support for the scenario proposed some years ago (11-14) for the phase diagram of cuprate superconductors where generalized Wigner

A. Bianconi (ed.), Symmetry and Heterogeneity in High Temperature Superconductors, 147–156.

polaron crystals are formed at critical values of the charge densities, i.e., at integer values of the filling factor.

The critical densities for the formation of the electronic polaronic crystals depend on the effective volume occupied by the doped charge carriers. The anisotropic interactions between the doped charge carriers produce the stripe or checkboard phases [15-16].

The van der Waals scheme is the simplest scheme to describe the spinodal phase separation that has been used to describe the co-existence of a polaronic gas (low density insulating phase) and a polaronic liquid (high density metallic phase) by Emin [1], or for more complex phase separation models including magnetic interactions [2-3].

In order to give account of the new emerging complex phase diagram of cuprates, where different phases coexist, we have extended the model of Poole et al. [17] for supercooled water.

2. THE MODEL

We use a modified Van der Waals interaction model, analogously to the one introduced for the phase diagram of supercooled water [17]. In fact, supercooled water, a prototype of complex matter, shows a phase separation driven by the tendency of water molecules to form arrays of hydrogen bonds. This tendency gives fluctuating clusters made of a low density liquid (LDL) that coexists with the high density liquid water (HDL). In order to describe this anomalous phase separation in water, the model of Poole et al. [47] implements the standard Van der Waals model by including a characteristic gain in energy "γ" for the formation of clusters with directional hydrogen bonds at a particular preferential volume "V_γ". The introduction of this anisotropic interaction provides a phase diagram with the coexistence of a high density liquid (HDL) and a low density liquid (LDL) when γ is larger than a threshold value [47].

Here we have extended the Poole model in order to describe multiple phase separations in cuprates. The free energy of a complex system in which n phase separations are observed, is obtained by adding the term A_γ, to the Van der Waals free energy A_{VDW} yielding a total free energy

$$A = A_{VDW} + A_\gamma \tag{1}$$

where A_{VDW} and A_γ are given by

$$A_{VDW} = -RT\left\{\ln\left[{V-b}/{\Lambda^3}\right]+1\right\}-{a}/{V^2} \quad (2)$$

$$A_\gamma = -RT\sum_{i=1}^{n-1} f_i \ln\left[\Omega_i + \exp(-\gamma_i/RT)\right] - \left[1 - RT\sum_{i=1}^{n-1} f_i \ln(\Omega_i + 1)\right] \quad (3)$$

Here a and b are the standard Van der Waals constants. The Van der Waals constant a is associated with the isotropic inter-particle attraction. Each intermediate phase, i , is characterized by the anisotropic inter-particle interaction γ_i. In this approach, for each phase i there are Ω_i>>1 configurations all having $\gamma_i = 0$ and only a single configuration in which the formation of the anisotropic bonds with energy γ_i is allowed.

The anisotropic interaction, characteristic of the intermediate phase *i,* is most likely to occur when the bulk molar volume is consistent with a preferential volume $V_{\gamma i}$. In fact, each particle has an optimal local volume for the formation of anisotropic bonds to its neighbours. Changing the particle density, when $V \neq V_{\gamma i}$, the anisotropic interactions are only a fraction f_i of the total, since V is no longer consistent with the possibility that all anisotropic interactions are saturated at the optimal volume. The remaining fraction of bonds, $1-f_i$, occurs in an unfavourable local volume and therefore, they cannot form the anisotropic bonds of the *i* phase.
The term f_i is given by

$$f_i = \exp\left\{-[(V - V_{\gamma i})/\sigma_i]^2\right\} \quad (4)$$

where σ_i characterizes the width of the region of volume around $V_{\gamma i}$ over which a significant fraction of anisotropic bonds can be described by Eq. (3).

To describe the phase diagram of cuprates, where we observe four phase separations, we consider a Poole model with three different anisotropic interactions γ_1, γ_2, and γ_3 giving rise to three different intermediate phases with preferential volume $V_{\gamma 1}$, $V_{\gamma 2}$ and $V_{\gamma 3}$ respectively.

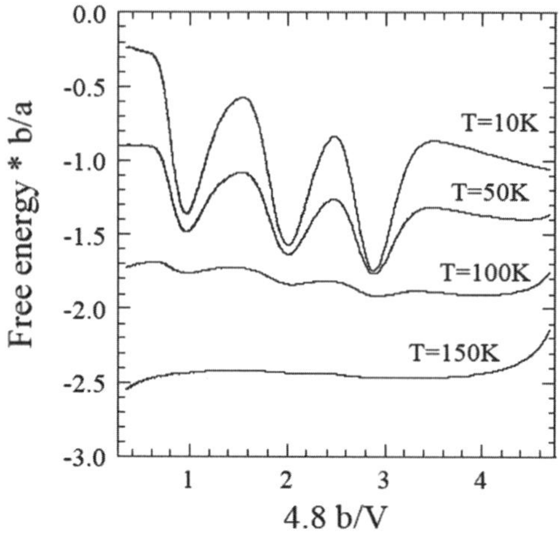

Figure III:4:1. The Helmotz free energy as a function of reduced density at different temperatures. In this way the minima of the free energy in Fig. 1 occur at the reduced density, or filling factor 1, 2 and 3. These new minima, indicating the occurrence of three intermediate phases at different densities, become deeper when the temperature decreases, as shown in Fig. 1 for T=10K, 50K, 100K and 150K. The occurrence of these new intermediate phases is also shown in Fig. 2 where we have plotted the state equation with pressure P as a function of reduced density, at the same temperatures T = 10K, 50K, 100K and 150K.

The values of the parameters $a, \sigma_i, \gamma_i, \Omega_i$ are fixed to get a qualitative agreement with experimental data. For the standard Van der Waals model we have chosen a=1.2 Pa m^6/mol^2, with $\gamma_0 = a/b$=6 KJ/mol =62 meV and b=2.00·10^{-4} m^3/mol. We have selected b in such a way that the particle effective volume is a sphere with radius r=4.3 Å. For the first intermediate phase we use $|\gamma_1|$=6.75 KJ/mol=70 meV, $V_{\gamma 1}$=9.35·10^{-4} m^3/mol and $\sigma_1 = V_\gamma/4$. For the second intermediate phase we use $|\gamma_2|$=6.75 KJ/mol=70 meV, $V_{\gamma 2}$=4.81·10^{-4} m^3/mol and $\sigma_2 = V_\gamma/7$, while for the third intermediate phase we impose $|\gamma_3|$=6.75 KJ/mol=70 meV, $V_{\gamma 3}$=3.34·10^{-4} m^3/mol and $\sigma_3 = V_\gamma/11$. In all intermediate phases we have chosen $\Omega_1 = \Omega_2 = \Omega_3 = exp(-S_\gamma/R)$ where S_γ =-70.25 J/(K mol) is the entropy of formation of a mole of anisotropic bonds and R is the universal gas constant.

The inclusion in the model of three optimum volumes $V_{\gamma 1}$ $V_{\gamma 2}$ and $V_{\gamma 3}$ for three different anisotropic bonds γ_1, γ_2 and γ_3, introduces three new minima in the free energy when V approaches $V_{\gamma 1}$ $V_{\gamma 2}$ and $V_{\gamma 3}$. In Fig. 1 we have plotted the free energy as a function of the reduced density in units of 1/4.8b.

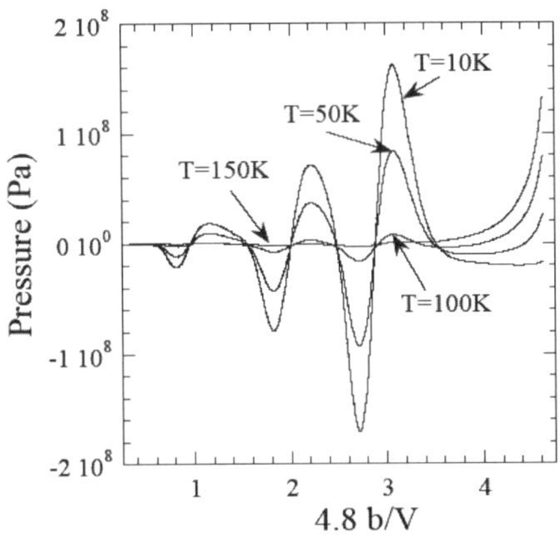

Figure III:4:2. The pressure as a function of the reduced density at different temperatures. The phase diagrams, obtained by calculating the spinodal line from Eq. 1, depend on the thermodynamic parameters described above. In Fig. 3 we report same representative phase diagrams with different anisotropic interactions

In panel a), we have used the values above $\gamma_0=a/b=62$ meV, $|\gamma_1|=|\gamma_2|=|\gamma_3|=70$ meV. We observe that the effect of the A_γ term in Eq. 1, related to the strength of the anisotropic interactions, is to "split" the normal Van der Waals spinodal curve by imposing thermodynamic stability in the region of states centered at the reduced density 1, 2 and 3 where the three different intermediate phases are stable. As a result, four spinodal lines occur, each terminating at a critical point producing four phase separation regions. As the directional bond energies $|\gamma_i|$ (i>1) decrease respect with the Van der Waals interaction$|\gamma_0|$, the phase separations generated by the strength of the directional bonds decrease and the stabilizing effect of $A_{\gamma i}$ set in only at lower temperature and the critical points C_n merge with the high density spinodal of the main Van der Waals spinodal line that is formed when all $|\gamma_i|$ go to zero (panel f).

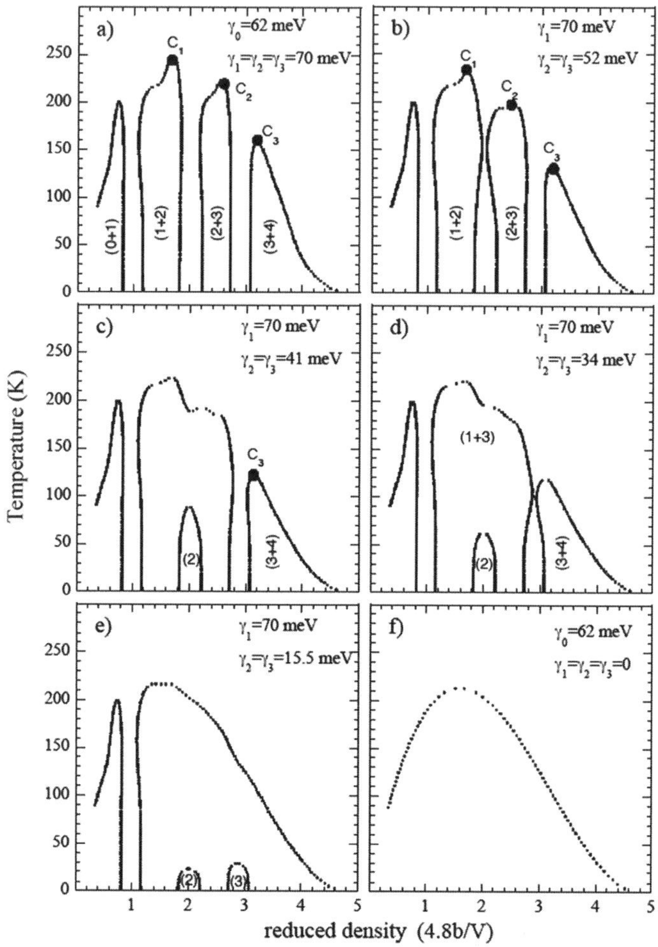

Figure III:4:3. The phase diagrams obtained by computing the spinodal lines from Eq. 1. We can observe the occurrence of four phase separations indicated by the four spinodal lines in panel a). In the panel b), c), d), e) and f) we show the effect of the strength of the directional bonds (see text)

In the following phase diagrams of panel b), c), d), e) we keep the γ_1 value fixed (γ_1=70 meV), while we change the γ_2 and γ_3 values in order to show the effects of the strength of the anisotropic interactions on the thermodynamic behaviour of the system.

In panel b) we observe that the two spinodal lines, (1+2) and (2+3), are going to overlap decreasing γ_2 and γ_3 ($\gamma_2 = \gamma_3$=52 meV) so that the phase separation due to the thermodynamic stability centred at reduced density 2 is going to disappear below a critical value of the interaction $|\gamma_2|=|\gamma_3|\sim$50meV. The disappearance of this phase separation is clearly shown in panel c) where decreasing the γ_2 and γ_3 interactions ($\gamma_2= \gamma_3$=41 meV) the intermediate low density phase centred at reduced density ρ=2, becomes completely enclosed within a spinodal line in which the states would otherwise be unstable. On the other hand the third intermediate phase, centred at reduced density ρ=3 is stable and the phase separation (3+4) at high density is still present. If we continue to decrease the γ_2 and γ_3 interactions we reach the critical threshold $\gamma_2= \gamma_3$=34 meV, where also the third phase separation (3+4) merges with the one at lower density as shown in the panel d). When γ_3 becomes less than 34 meV the third intermediate low density phase centred at reduced density r=3 becomes an isolated pocket of stability completely enclosed within a spinodal line as shown in panel e), where we have used $\gamma_2= \gamma_3$=15.5meV. In this case the action due to the central and symmetric interaction *a* is prevalent since $|\gamma|$ is lower than a characteristic value, and the phase separations between the separate spinodal curves becomes two isolate "pockets" of stability at the reduced density 2 and 3 below the low temperatures 22K and 29K respectively. Finally, when the anisotropic interactions γ_1, γ_2 and γ_3, go to zero we reach a quantum critical point where we obtain the standard Van der Waals phase diagram shown in panel f).

3. THERMODYNAMIC BEHAVIOUR OF COMPLEX OXIDES

Now we apply this model to the $La_{2-x}Sr_xCuO_4$ (LSCO) and $Bi_2Sr_2CaCu_2O_{8+\delta}$ (Bi2212) compounds.

We report in Fig.4 the pseudo-gap temperatures T* measured by Singer et al. [18] with the pseudo-gap temperature T* in EXAFS experiments [19].

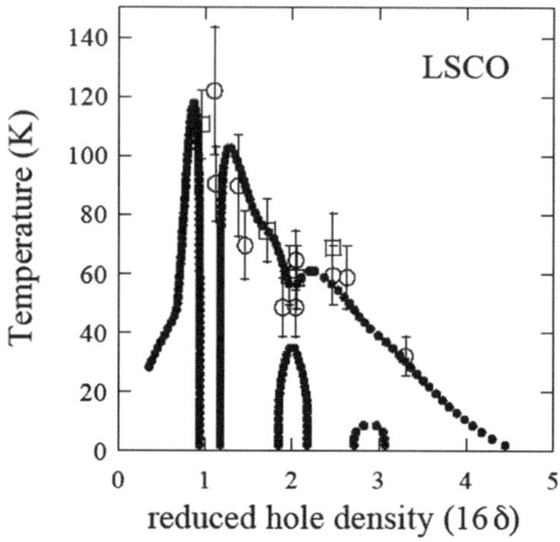

Figure III:4:4. The phase diagram for the LSCO: the open circles correspond to the T* temperatures obtained by Singer [18]; the squares indicate the values of T* measured by X ANES spectroscopy by our group

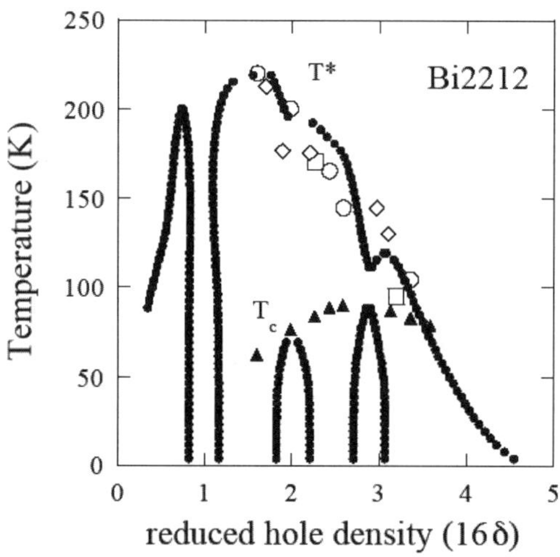

Figure III:4:5. The phase diagram for the Bi2212: the circles and the squares correspond to the T* temperatures obtained from experiments by Oda [20] and from Arpes measurements by Ding [21] respectively; the diamonds indicate the EXAFS T* measurements by Saini et al. [22]. The superconducting T_c are also indicated by the full triangles

We have reproduced the experimental pseudo-gap temperature using the thermodynamic model above discussed. The phase diagram has been obtained calculating the spinodal line from Eq.1. We have imposed $\gamma_0=a/b=19.5$ meV, $b=2.00\cdot10^{-4}$ m^3/mol, $|\gamma_1|=44$meV, $V_{\gamma 1}=9.25\cdot10^{-4}$ m^3/mol, $\sigma_1=V_\gamma/6.25$, $|\gamma_2|=14$ meV, $V_{\gamma 2}=4.81\cdot10^{-4}$ m^3/mol, $\sigma_2=V_\gamma/8.33$, $|\gamma_3|=5.2$ meV, $V_{\gamma 3}=3.34\cdot10^{-4}$ m^3/mol, $\sigma_3=V_\gamma/11$ and $\Omega_1=\Omega_2=\Omega_3=exp(-S_\gamma/R)$ where $S_\gamma=-70.25$ $J/(K\ mol)$.

In Fig. 5 we report the experimental pseudo-gap temperatures T* for the Bi2212 system [20-21]. The calculated phase diagram has been obtained by imposing $\gamma_0=a/b=62$ meV, $b=2.00\cdot10^{-4}$ m^3/mol, $|\gamma_1|=70$meV, $V_{\gamma 1}=1.00\cdot10^{-3}$ m^3/mol, $\sigma_1=V_\gamma/6.25$, $|\gamma_2|=37$ meV, $V_{\gamma 2}=4.81\cdot10^{-4}$ m^3/mol, $\sigma_2=V_\gamma/8.33$, $|\gamma_3|=33$ meV, $V_{\gamma 3}=3.34\cdot10^{-4}$ m^3/mol, $\sigma_3=V_\gamma/11$ and $\Omega_1=\Omega_2=\Omega_3=exp(-S_\gamma/R)$ where $S_\gamma=-70.25$ $J/(K\ mol)$.

In conclusion we have presented a model for an electronic complex system with coexistence of different electronic phases at critical densities and coexistence of different liquids described by the modified van der Waals model as proposed for supercooled water. We discuss the critical values of the anisotropic interactions for the spinodal lines. We find that this model is able to describe the evolution of the pseudo-gap temperature versus doping in different cuprate families.

AKNOWLEDGEMENTS

This work is supported by MIUR in the frame of the project Cofin 2003 "Leghe e composti intermetallici: stabilità termodinamica, proprietà fisiche e reattività" on the "synthesis and properties of new borides."

REFERENCES

1. D. Emin, *Phys. Rev. B*, **49**, 9157 (1994); *Phys. Rev. Lett.* **72**, 1052 (1994).
2. J. C. Phillips, and J. Jung, *Phil. Mag. B* **81**, 745 (2001).

3. L.P. Gorkov in *Proc. of the Toshiba Intern. School of superconductivity* Kyoto Japan 15-20 July (1991). Springer Series in Solid State Sciences, S. Maekawa and M. Sato editors vol. 106 pag.71 Springer Verlag Berlin 1992.
4. K. A. Müller, G.-M. Zhao, K. Conder and H. Keller *Journal of Physics: Condensed Matter* **10**, L291-L296 (1998); D. Mihailovic and K. A. Müller in *"High-T_C Superconductivity 1996: Ten Years after the Discovery"*, edited by E. Kaldis, E. Liarokapis and K. A. Müller (Kluwer Academic, Dordrecht 1996) NATO ASI Series vol. **343** pp. 243.
5. *Phase Separation in Cuprate Superconductors* edited by K. A. Müller and G. Benedek (World Scientific, Singapore, 1993). " (Proc. of the workshop held in Erice, Italy, 6-12 May, 1992).
6. *"Phase Separation in Cuprate Superconductors"* edited by E. Sigmund and A. K. Müller (Springer Verlag, Berlin-Heidelberg, 1994), (proc. of the second int. workshop on Phase Separation held in Cottbus, Germany, Sept 4-10, 1993)
7. V. Cataudella, G. De Filippis, G. Iadonisi, A. Bianconi and N. L. Saini, *Int. Jour. of Modern Physics* B **14**, 3398 (2000).
8. A. Bianconi, G. Bianconi, S. Caprara, D. Di Castro, H Oyanagi, and N. L. Saini, *J. Phys.: Condens. Matter,* **12** 10655 (2000); A. Bianconi, N. L. Saini, S. Agrestini, D. Di Castro, and G. Bianconi *Int. Jour. of Modern Physics B* **14**, 3342 (2000).
9. T. Hanaguri, C. Lupien, Y. Kohsaka, D.-H. Lee, M. Azuma, M. Takano, H. Takagi, and J. C. Davis, *Nature* **430**, 1001 (2004).
10. K.M. Shen, F. Ronning, D.H. Lu, F. Baumberger, N.J.C. Ingle, W.S. Lee, W. Meevasana, Y. Kohsaka, M. Azuma, M. Takano, H. Takagi, Z.X. Shen, *Science* 307, 901 (2005).
11. A. Bianconi *Sol. State Commun.* 91, 1 (1994).
12. A. Bianconi, M. Missori *Sol. State Commun.* 91, 287 (1994).
13. A. Bianconi. *Physica C* 235-240, 269 (1994).
14. A. Bianconi, M. Missori, H. Oyanagi, and H. Yamaguchi D. H. Ha, Y. Nishiara and S. Della Longa, *Europhysics Letters* 31, 411 (1995).
15. F. V. Kusmartsev *Phys. Rev. Lett.* **84**, 530 (2000); *ibidem* **84**, 5026 (2000). F. V. Kusmartsev, *Europhys. Lett.* **54**, 786 (2001).
16. Anup Mishra, Michael Ma, Fu-Chun Zhang, Siegfried Guertler, Lei-Han Tang, and Shaolong Wan, *Phys. Rev. Lett.* **93**, 207201 (2004).
17. P. H. Poole, F. Sciortino, T. Grande, H. E. Stanley, and C. A. Angell, *Phys. Rev. Lett.* **73**, 1632 (1994) and references therein.
18. P. M. Singer, A. W. Hunt, A. F. Cederstro, and T. Imai, *Phys. Rev. B* **60**, 15345 (1999).
19. N. L. Saini, H. Oyanagi, Z. Wu and A. Bianconi, *Supercond. Sci. Technol.* **15**, (2002).
20. N. Momono, R. Dipasupil, G. Ishiguro, S. Saigo, T. Nakamo, M. Oda, and M. Ido, *Physica C* **317-318**, 603 (1999).
21. H. Ding, T. Yokoya, J. C. Campuzano, T. Takahashi, M. Randeria, M. R. Norman, T. Mochiku, K. Kadowaki, J. Giapintzakis, Nature (London) **382**, 6586 (1996).

III.5

ANISOTROPY OF THE CRITICAL CURRENT DENSITY IN HIGH QUALITY $YBa_2Cu_3O_{7-\delta}$ THIN FILM

A. Taoufik[1], A. Tirbiyine[1], A. Ramzi[1] and S. Senoussi[2]

[1]*Laboratoire des Matériaux Supraconducteurs à Haute Température Critique, Département de Physique, Faculté des Sciences, Université Ibn Zohr, B. P:8106, Agadir, Morocco.*

[2] *Laboratoire de Physique des Solides (associé au CNRS. URA. 0002), Université Paris Sud, Bâtiment 510, 91405 Orsay Cedex, France.*

Abstract: We have investigated the transport critical current density J_c as a function of the angle θ between the crystallographic *c*-axis and the applied magnetic field in high quality $YBa_2Cu_3O_{7-\delta}$ thin film. Measurements were performed for various temperature and magnetic field values. Our results show that the critical current density maximum occurs when the applied magnetic field is parallel to the *ab* planes ($\theta = 90°$).The angular dependence of the critical current density shows the existence of the intrinsic pinning between the CuO_2 layers for H parallel to the *ab* planes and the extrinsic pinning in the configuration where the magnetic field is parallel to the *c*-axis. We have analyzed our results in the framework of the intrinsic pinning model proposed by Tachiki and Takahachi

Key words: The Critical Current Density; Intrinsic Pinning; Anisotropy

A. Bianconi (ed.), Symmetry and Heterogeneity in High Temperature Superconductors, 157–162.

1. INTRODUCTION

The high critical temperature superconductors show a strong anisotropy in different properties: critical current density [1], resistivity [2, 3] and the upper critical field [2].

A large anisotropy in the superconducting critical current density J_c of high temperature superconducting thin film have been reported [4, 5]. A large enhancement when the applied magnetic field H (perpendicular to J) is precisely parallel to the copper-oxygen planes of the lattice structure. Also there are theoretical arguments that the enhancements arise from intrinsic flux pinning when the vortex cores are located in the weak superconducting regions between the Cu-O layers [6].

An intrinsic pinning model was proposed by Tachiki and Takahachi [7]. They derived the angular dependence of J_c. The intrinsic pinning model is based on the layer structure in the oxide superconductors which consist of strong superconducting layers such as CuO_2 and weak superconducting layers such as CuO chains and BaO planes in $YBa_2Cu_3O_{7-\delta}$. They supposed that weak superconducting layers work as natural pinning centers. The existence of extrinsic pinning centers, however was also assumed in this model.

They assumed that the flux lines may be pinned by weak superconducting layers (intrinsic pinning centers) or extrinsic pinning such as twin planes. They supposed that weak superconducting layers and twin planes work as pinning centers most effectively when $\theta = 90°$ and $\theta = 0$, respectively.

In this work, we have measured the critical current density of $YBa_2Cu_3O_{7-\delta}$ thin film in strong magnetic field H up to 10 T at various angle θ between H and the crystallographic c-axis. We have compared our results to the intrinsic pinning model proposed by Tachiki and Takahachi [7].

2. EXPERIMENTAL DETAILS

The epitaxial $YBa_2Cu_3O_{7-\delta}$ thin films were deposited by laser ablation method onto (100) surface of single crystal $SrTiO_3$. The sample showed a zero resistivity at 90 K in zero magnetic field. The film thickness and width were 400 nm and 7.53 μm, respectively. Electrodes of power measurement are in gold and deposited by in situ evaporation. The distance which separates this electrode was 135 μm. Contact resistance's were less than 1 Ω. Measurements were realized by using the DC four-probe method.

In order to rule out distortions of the *E-J* curve by extensive heating that could be induced by the very high extensive heating that could be induced by the very high dissipation levels employed here, a pulsed current power supply was used with a time duration τ = 10 ms, a waveform repeat time of 2 s and an average over 64 pulses at the same fixed *J*, *T* and *H*.

The microstructure of several of these thin films was studied extensively, using Transmission Electron Microscopy (T.E.M.). These T.E.M. observations together with X-ray Energy Dispersion Spectroscopy (E.D.S.) as well as usual X-ray spectra show that the films are highly homogeneous and have essentially a single YBCO (123) phase.

Transmission electron microscopy (T.E.M.) observations performed on our samples revealed not only the presence of the usual twin boundaries as the major visible defect but also, a set of columnar-like defects. In addition, the sample certainly contains also point defects, in particular oxygen vacancies.

3. RESULTS AND DISCUSSION

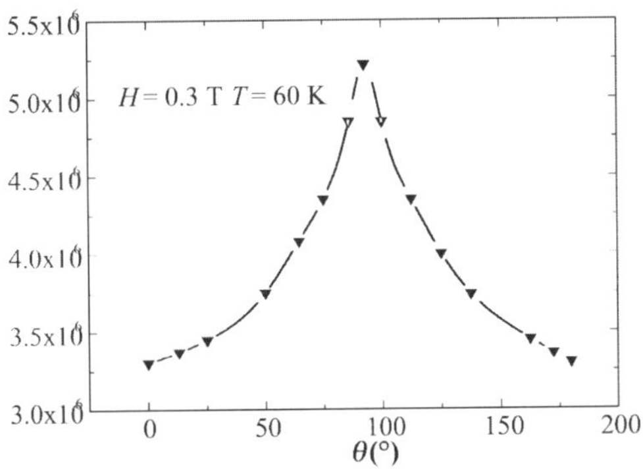

Figure III:5:1. The critical current density as a function of the angle θ at 0.3 T and 60 K

In figure 1, we present an example of the critical current density J_c variations as a function of the angle θ between the crystallographic *c*-axis and the applied magnetic field.

The magnetic field and temperature values are 0.3 T and 60 K, respectively.

As can be seen in this figure, J_c increases as θ increases, it reaches its maximal value at $\theta = 90°$ corresponding to the configuration where the applied magnetic field is adjusted parallely to the *ab* planes of the sample. After the maximum value, $J_c(\theta)$ decreases and reaches its initial value.

We plot in figure 2 the angular dependence of the critical current density J_c/J_{c1}, J_{c1} is the critical current density when a magnetic field is applied parallel to the film surface. We note that J_{c1} is independent of the angle θ. The solid circles are the experimental data obtained for $YBa_2Cu_3O_{7-\delta}$ thin film at 10 T and 60 K. For comparison, the solid curve presents the theoretical values given by [8]

$$J_c = \frac{J_{c2}}{\left|\cos\theta\right|^{1/2}}$$

where J_{c2} is the critical current density in the configuration where the magnetic field is parallel to the *c*-axis. The values of J_{c1} and J_{c2} are 1.35×10^6 A/cm^2 and 3.7×10^5 A/cm^2, respectively.

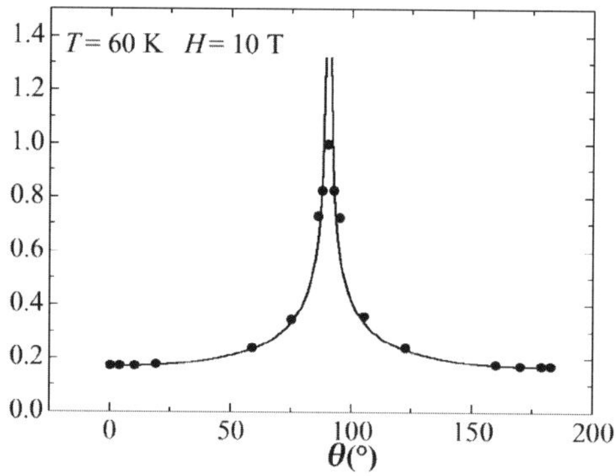

Figure III:5:2. Angular dependence of J_c/J_{c1} at 60 K and 10 T as a function of the angle θ. $J_{c1} = 1.35\times10^6$ A/cm^2 and $J_{c2} = 3.7\times10^5$ A/cm^2

As can be seen, in this experimental condition, the theoretical curve proposed by the intrinsic pinning model of Tachiki and Takahachi are in good agreement with our experimental data.

The experimental values obtained at 60 K, in an applied magnetic field of 0.6 T are plotted in figure 3. For comparison, with the theoretical values the solid line presents the intrinsic pinning model values using $J_{c1} = 6.07\text{x}10^6$ A/cm^2 and $J_{c2} = 2.3\text{x}10^6$ A/cm^2.

The concordance between the two curves is less good than in the case where the applied magnetic field is 10 T especially for angles close of 90°.

A single crystal film of $YBa_2Cu_3O_{7-\delta}$ is considered with the CuO_2 layers parallel to the film surface. These layers and their vicinities are strongly superconductive and the other layers like CuO chains are weakly superconductive. Accordingly, these crystals are considered to be constructed by an alternate stacking of strongly and weakly superconducting layers.

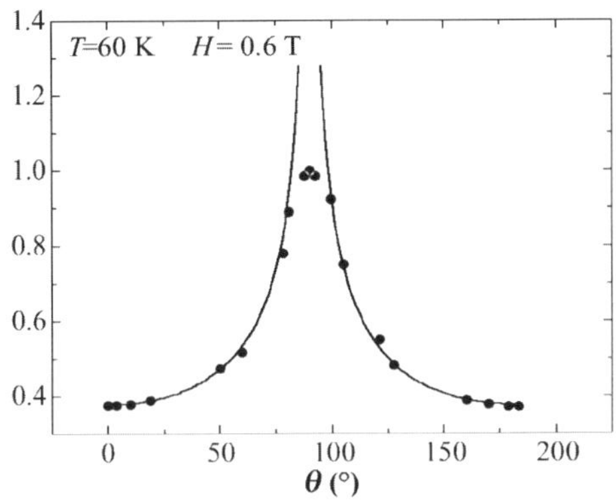

Figure III:5:3. Angular dependence of J_c/J_{c1} at 60 K and 0.6 T as a function of the angle θ. $J_{c1} = 6.07\text{x}10^6$ A/cm^2 and $J_{c2} = 2.3\text{x}10^6$ A/cm^2

At low temperature, flux lines preferentially penetrate into the weakly superconducting layers cause they are stabilized the most when they are at the weakly superconducting layers, since the loss of the superconducting energy due to the inclusion of the flux lines is least in this case.

The peak in $J_c(\theta)$ at $\theta = 90°$ corresponding to the configuration where the magnetic field is parallel to the *ab* planes can be explained by the following pinning mechanism which is characteristic for the layered oxide. The weakly superconducting layers work as natural pinning centers. The pinning strength is considerably high, and thus the critical current density becomes very high. The critical current density is strongly dependent on the direction of the applied magnetic field.

In the weak magnetic field region (0.6 T), the experimental results are not in agreement with the intrinsic pinning model. In this case, the flux pinning at the film surface becomes more dominant than the intrinsic pinning by the *ab* planes.

We have reported a large anisotropy in the critical current density J_c of high quality $YBa_2Cu_3O_{7-\delta}$ thin films. A large enhancement of J_c when the applied magnetic field H is precisely parallel to the Cooper-Oxygen planes of the lattice was observed.

REFERENCES

1. Y. Enomoto, T. Murakami, M. Suzki, K. Moriwaki, Jpn. J. Appl. Phys., 1987; 26: L1248.
2. Y. Iye, T. Tamegar, T. Sakakibara, T. Goto, N. Miura, H. Takeya, H. Takei, Physica C, 1988; 153-155: 26.
3. A. Taoufik, S. Senoussi, A. Tirbiyine, Ann. Chim. Sci. Mat, 1999; 24: 227-232.
4. D. K . Christen, et al., Physica C, 1989; 162: 653.
5. B. Roas, L. Schultz, G. Saemann-Ischenko, Phys. Rev. Lett., 1990; 64: 479.
6. M. Tachiki, S. Takahachi, Physica C, 1989; 162-164: 241.
7. M. Tachiki, S. Takahachi, Solid State Commun., 1989; 70: 291.
8. M. Tachiki, S. Takahachi, Solid State Commun., 1989; 72: 1083-1086.

IV

SYMMETRY OF THE CONDENSATE

IV.1

SYMMETRY OF HIGH-T_c SUPERCONDUCTORS

F. Iachello
Center for Theoretical Physics, Sloane Physics Laboratory, Yale University, New Haven, CT 06520-8120

Abstract: The symmetry classification of superconducting states is reviewed. Based on purely symmetry considerations, a simple proof is given for the enhancement of $d_{x^2-y^2}$ superconductivity at the surface of cuprate materials. A novel method to study mixed superconducting phases is introduced.

1. INTRODUCTION

In the last two decades, important discoveries in the field of superconductivity [1] have reopened the question of what is the symmetry of the superconducting state. In this contribution, after a brief historical introduction, the classification of superconducting states will be reviewed. Some consequences of the symmetry of the state will be then discussed, especially in view of the recent interest in the symmetry of high-T_c superconductors. Finally, a novel approach particularly useful for mixed symmetry states will be introduced.

In 1957, Bardeen, Cooper and Schriffer proposed a microscopic theory of superconductivity [2]. This theory was isotropic (s-wave pairing). In 1958, Bogoliubov [3] and Valatin [4] introduced a transformation that made the treatment of superconductivity simpler (quasi-particle transformation). Still in 1958, Anderson [5] addressed the same problem by introducing the algebra of SU(2) to describe the properties of the system (quasi-spin algebra). In the same year, Bohr, Mottelson and Pines applied BCS theory to

A. Bianconi (ed.), Symmetry and Heterogeneity in High Temperature Superconductors, 165–180.

the study of atomic nuclei and found evidence for s-wave pairing in nuclei [6]. The symmetry of the superconducting state was enlarged by Anderson and Morel in 1961 [7] and by Balian and Werthamer in 1963 [8] (p-wave pairing). Up to the late 60's, BCS theory based on s-wave pairing was used both in condensed matter systems and in atomic nuclei. However, beginning in the early 70's it become apparent that other types of symmetry play a role. It is convenient at this stage to separate rotational invariant systems from point group invariant systems. For rotationally invariant systems, the symmetry group to be described in Sect.2 is $G \equiv O(3)$. In 1973 Leggett [9] proposed p-wave pairing to describe properties of ^3He and in 1974 Arima and Iachello [10] proposed a combination of s- and d-wave pairing to describe properties of atomic nuclei. For point group invariant systems, the symmetry group is $G \equiv$ *Point group*. In 1979, Steglich et al. [11] discovered superconductivity in heavy fermion materials. Some of the point groups relevant to these materials are D_{4h}, O_h and D_{6h}. Finally, in 1986 the discovery of high-T_c superconductivity in cuprate materials with point group D_{4h} by Bednorz and Müller [1] brought the question of the symmetry of the superconducting state back to the forefront of physics.

An excellent account of the symmetry of superconducting states is given in the review article of Sigrist and Ueda [12]. In the first part of this contribution, this article will be followed and consequences of symmetry on the surface of cuprate materials will be derived. In the second part of the contribution, a novel approach to the study of mixed superconducting phases will be introduced and its connection to the Ginzburg-Landau approach briefly discussed.

2. PAIRING HAMILTONIAN

The starting point for the study of the symmetry of the superconducting state is the pairing Hamiltonian. For applications to condensed matter systems it is convenient to write this Hamiltonian in momentum space

$$\begin{aligned} H = & \sum_{\vec{k},s} \varepsilon(\vec{k}) a^\dagger_{\vec{k},s} a_{\vec{k},s} \\ & + \frac{1}{2} \sum_{\vec{k},\vec{k}',s_1,s_2,s_3,s_4} \upsilon_{s_1,s_2,s_3,s_4}(\vec{k},\vec{k}') a^\dagger_{-\vec{k},s_1} a^\dagger_{\vec{k},s_2} a_{-\vec{k}',s_3} a_{-\vec{k}',s_4} \end{aligned} \quad . \tag{1}$$

Here $\varepsilon(\vec{k})$ is the single particle energy and $\upsilon(\vec{k},\vec{k}')$ the interaction. The symmetry of the superconducting state can be derived from that of the Hamiltonian. In general, the symmetry group $\mathcal{G}$ is the direct product

$$\mathcal{G} \equiv G \otimes SU_s(2) \otimes \mathcal{T} \otimes U(1) \tag{2}$$

where for point group invariant systems G = *Point group*, while for rotational invariant systems $G = O(3)$. The group $SU_s(2)$ describes ordinary spin, while $\mathcal{T}$ is the time-reversal group and $U(1)$ the gauge group. In this article, the discussion will be limited to G, which will be called the intrinsic group.

By introducing gap functions $\Delta_{S_1S_2}(\vec{k})$, H can be approximated by a one-particle Hamiltonian

$$\begin{aligned}\tilde{H} = & \sum_{\vec{k},s} \varepsilon\left(\vec{k}\right) a^{\dagger}_{\vec{k},s} a_{\vec{k},s} \\ & + \frac{1}{2} \sum_{\vec{k},s_1,s_2} [\Delta_{s_1,s_2}\left(\vec{k}\right) a^{\dagger}_{\vec{k},s_1} a^{\dagger}_{-\vec{k},s_2} - \Delta^{*}_{s_1,s_2}\left(-\vec{k}\right) a_{-\vec{k},s_1} a_{\vec{k},s_2}].\end{aligned} \tag{3}$$

The Bogoliubov transformation

$$a_{\vec{k},s} = \sum_{s'} \left(u_{\vec{k},ss'} \alpha_{\vec{k},s'} + \upsilon_{\vec{k},ss'} \alpha^{\dagger}_{-\vec{k},s'} \right) \tag{4}$$

brings this Hamiltonian in an even simpler form, in terms of quasi-particle operators $\alpha_{\vec{k},s}$ and quasi-particle energies, $E_{\vec{k},s}$.

3. CLASSIFICATION OF SUPERCONDUCTING STATES

A complete classification of superconducting states is given in the review paper of Sigrist and Ueda [12], both for triplet and singlet pairing. Since it appears that in high-T_c superconductors one has singlet pairing, only this case will be discussed in this article. We are interested here in obtaining a basis for the representations of the group $\mathcal{G}$ of Eq.(2).

For rotational invariant systems, the group $G \equiv O(3) = SO(3) \otimes \mathcal{P}$,where $\mathcal{P}$ is the parity operation. Leaving aside time-reversal and gauge groups and noting that $S = 0$ (singlet states), we are led to consider the classification of the representations of $O(3)$. These are labeled by the integer number $\ell = 0, 1, 2,...$ The parity is $(-)^{\ell}$ and can be omitted.
We thus have

$$D(\mathcal{G}) = D_\ell(G) \otimes D_S(SU_s(2)) \otimes D_{\mathcal{T}}(\mathcal{T}) \otimes D_\varphi(U(1)) \tag{5}$$

The basis functions for the intrinsic group G are the polynomial harmonics in momentum space,

$$k^\ell Y_{\ell m}\left(\hat{k}\right). \tag{6}$$

[The spherical harmonics $Y_{\ell m}(\hat{k})$ are a basis for the representations of $SO(3)$. Here the polynomial harmonics are used to construct the basis states in the following tables.] For singlet pairing, only positive parity harmonics are of importance. The lowest positive parity harmonics are given in Table 1 (real form). For studying surface phenomena, we also need negative parity harmonics. The lowest negative parity harmonics are given in Table 2 (real form).

Table IV:1:1. Real forms of the lowest ($\ell \leq 4$) positive parity harmonics. $P_m^\ell(\theta)$ are the associated Legendre polynomials

$\ell = 0$	P_0^0
$\ell = 2$	$k^2 P_m^2(\theta)\begin{cases}\cos m\varphi \\ \sin m\varphi\end{cases}$
$\ell = 4$	$k^4 P_m^4(\theta)\begin{cases}\cos m\varphi \\ \sin m\varphi\end{cases}$

Table IV:1:2. Real forms of the lowest ($\ell \leq 3$) negative parity harmonics

$\ell = 1$	$kP_m^1(\theta)\begin{cases}\cos m\varphi \\ \sin m\varphi\end{cases}$
$\ell = 3$	$k^3 P_m^3(\theta)\begin{cases}\cos m\varphi \\ \sin m\varphi\end{cases}$

For point group invariant systems, the intrinsic group is G = *Point group*. The construction of the basis for these systems is a standard group theoretical problem. For the groups D_{4h}, D_{6h} and O_h it was done by Hamermesh many years ago [13]. I report here only the case of $G \equiv D_{4h}$. For positive parity one has Table 3. For negative parity one has Table 4.

The representations here are labelled by the group theoretical notation [13] A_1, A_2, B_1, B_2, E. The first four are one dimensional, while the representation E is two dimensional. For $\ell \geq 3$, some representations are contained twice and the situation is slightly more complicated. In condensed matter physics, it has become customary to label the representations with the letter Γ [14]. When both positive and negative parity states are considered also the parity label is added, Γ^+ and Γ^-. The conversion between the two notations is $A_1 \rightarrow \Gamma_1$, $A_2 \rightarrow \Gamma_2$, $B_1 \rightarrow \Gamma_3$, $B_2 \rightarrow \Gamma_4$, and $E \rightarrow \Gamma_5$.

For high-T_c superconductors it appears that only s- and d-wave pairing is important. It appears also that superconductivity is in the CuO planes (x-y plane). Restricting the classification to two-dimensions one has Table 5. Here in the last column, the notation often used in high-T_c superconductivity is also indicated. Also a star is placed on the representation Γ_1 originating from $\ell = 2$ to distinguish it from that originating from $\ell = 0$, although both transform in the same way under D_{4h}. Another notation is "extended s-wave".

Table IV:1:3. Construction of the positive parity basis ($\ell \leq 2$) for the group D_{4h}

$\ell = 0$	A_1	P_0^0	1
$\ell = 2$	A_1	$k^2 P_0^2(\theta)$	$k_z^2; k_x^2 + k_y^2$
	A_2	–	–
	B_1	$k^2 P_2^2(\theta)\sin 2\varphi$	$k_x^2 - k_y^2$
	B_2	$k^2 P_2^2(\theta)\cos 2\varphi$	$k_x k_y$
	E	$k^2 P_1^2(\theta)\begin{cases}\cos\varphi \\ \sin\varphi\end{cases}$	$\begin{cases}k_y k_z \\ k_z k_x\end{cases}$

Table IV:1:4. Construction of the negative parity basis ($\ell \leq 1$) for the group D_{4h}

$\ell=1$	A_1	–		–
	A_2	$kP_0^1(\vartheta)$		k_z
	B_1	–		–
	B_2	–		–
	E	$kP_1^1(\vartheta)$	$\begin{cases}\cos\varphi\\ \sin\varphi\end{cases}$	$\begin{cases}k_x\\ k_y\end{cases}$

4. CONSEQUENCES OF SYMMETRY: QUASI-PARTICLE SPECTRUM

An immediate consequence of symmetry is the nature of the quasi-particle spectrum. The density of states $\rho(\omega)$ can be calculated from the knowledge of the quasi-particle energies. For singlet states $E_{k,\pm} = E_k$, with

$$E_k = \left(\varepsilon^2(\vec{k}) + \Delta^2(\vec{k})\right)^{1/2} . \tag{7}$$

In ordinary superconductors (s-wave pairing), $\Delta(\vec{k}) = \Delta_0$, and

$$\rho(\omega) = \begin{Bmatrix} 0 & \omega < \Delta_0 \\ N(0) \quad \dfrac{\omega}{\sqrt{\omega^2 - \Delta_0^2}} & \omega > \Delta_0 \end{Bmatrix}. \tag{8}$$

As an example of unconventional pairing consider the polar state in p-wave pairing. The density of states is given here by

$$\rho(\omega) = \begin{Bmatrix} N(0)\frac{\pi}{2}\frac{\omega}{\Delta_0} & \omega < \Delta_0 \\ N(0) \; \frac{\omega}{\Delta_0}\arcsin\frac{\Delta_0}{\omega} & \omega > \Delta_0 \end{Bmatrix}. \tag{9}$$

In general, for rotational invariant systems, the gap function can be expanded into polynomial harmonics

$$\Delta(\vec{k}) = \Delta_0 \sum_{\ell m} c_{\ell m} k^{\ell} Y_{\ell m}(\hat{k}) \tag{10}$$

Table IV:1:5. Restriction of the basis to two-dimensions

$\ell = 0$	Γ_1	1	s
$\ell = 2$	Γ_{1*}	$k_x^2 + k_y^2$	s_*
	Γ_3	$k_x^2 - k_y^2$	$d_{x^2-y^2}$
	Γ_4	$k_x k_y$	d_{xy}

while, for point group invariant systems, it can be expanded into Γ_i's

$$\Delta(\vec{k}) = \Delta_0 \sum_i c_i \Gamma_i(\vec{k}) \tag{11}$$

In two dimensions, and point group D_{4h}, the expansion is

$$\Delta(\vec{k}) = \Delta_0 \left[c_1 \Gamma_1(\vec{k}) + c_3 \Gamma_3(\vec{k}) + c_4 \Gamma_4(\vec{k}) \right]. \tag{12}$$

It is convenient to introduce polar coordinates, k and φ, as in Fig. 1. The expansion becomes then, for k independent gaps, i.e. dropping the factor k^2 in Γ_3 and Γ_4,

$$\begin{aligned} \Delta(\varphi) &= \Delta_0 \left[c_1 + c_3 \cos 2\varphi + c_4 \sin 2\varphi \right] \\ &\equiv \Delta_0 f(\varphi) \\ 0 \le \varphi &\le \frac{\pi}{2}. \end{aligned} \tag{13}$$

If only $c_3 \neq 0$, the gap has a line of nodes at $\varphi = \pi / 4$. One can introduce the quantity

$$\rho(\omega,\varphi) = \begin{Bmatrix} 0 & \omega < \Delta_0 f(\varphi) \\ N(0) \dfrac{\omega}{\sqrt{\omega^2 - \Delta_0^2 f(\varphi)^2}} & \omega > \Delta_0 f(\varphi) \end{Bmatrix} \tag{14}$$

and obtain the total density of states by

$$\rho(\omega) = \frac{1}{\pi/2} \int_0^{\pi/2} \rho(\omega,\varphi) d\varphi. \tag{15}$$

The dependence of $\rho(\omega)$ as $\omega \to 0$ is reflected in the behavior of several physical quantities as a function of temperature.

Since the density of states depends on the symmetry of the gap, any measurement sensitive to $\rho(\omega)$ will give information on the symmetry of the gap. An example is ARPES measurements [15]. The photoemission intensity can be written as

$$I(\omega) = \int w(\omega - \omega')S(\omega')\rho(\omega')d\omega', \tag{16}$$

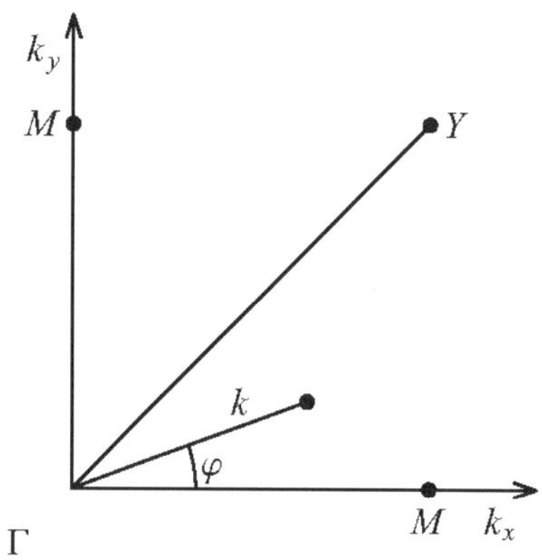

Figure IV:1:1. Choice of coordinates for expansion of the gap function

where the weight function depends on the energy resolution of the apparatus and can be taken as a Lorentzian

$$w(\omega - \omega') = \frac{\gamma^2}{(\omega - \omega')^2 + \gamma^2} \tag{17}$$

with width at half-maximum γ and $S(\omega)$ is the sensitivity of the instrument. Recent ARPES measurements [16] appear to indicate that in high-T_c superconductors the symmetry of the gap is $d_{x^2-y^2}$. However, ARPES measurements are sensitive only to the surface, of the order of 10× lattice constants. The question then arises on whether or not the symmetry of the gap is uniform throughout.

5. SURFACE VERSUS BULK

In 1974, Ambegaokar et al. [17] noted that anisotropic superconductivity is strongly influenced by a boundary within the range of its coherence length. They noted that in superfluid ^3He (p-wave superconductor) the internal angular momentum of the Cooper pairs always turns perpendicular to the vessel wall of the confining fluid. In an independent devolopment, in 1974 Arima and the author [10] introduced a model of atomic nuclei seen as liquid drops with s- and d-wave pairing. By analyzing the spectra of several atomic nuclei they concluded that some nuclei (Sn,Pb,...) are characterized by purely s-wave pairing while others (Sm,Gd,..;Pt,Os,...) are characterized by a mixture of s-and d-wave, with the bulk being only s-wave and the surface being a mixture of s- and d-wave, in fact mostly d-wave [18]. It thus appears that anisotropic pairing in rotational invariant systems with a surface is strongly influenced by the presence of a boundary. Recently, Müller has suggested that the same situation occurs in high T_c superconductors [19].

The arguments of Ambegaokar et al. are based on the introduction of a coordinate dependent gap

$$\Delta_{s_1 s_2}(\hat{k},\vec{r}) = \sum_m \eta_m(\vec{r})\Delta_m(\hat{k}), \tag{18}$$

where $\eta_m(\vec{r})$ are coordinate dependent order parameters, and on the solution of the appropriate integral equation. A much simpler argument can be given by pure symmetry considerations which are particularly appropriate for systems with point group symmetry. Consider a surface S with normal $\vec{n} \equiv (n_x, n_y, n_z)$ and couple the vector $\vec{n}$ to the order parameter. For rotational invariant systems, the normal vector $\vec{n}$ belongs to the vector representation $D_{\ell=1}(G)$. From Table 4 one can see that it belongs to the representations $\Gamma_2^- \oplus \Gamma_5^-$ of D_{4h}. When coupled to a representation Γ of the order parameter it induces terms of the type

$$D^d \otimes \Gamma^* \otimes \Gamma. \tag{19}$$

The power of D's must be even $d = 2, 4, \ldots$ since a vector has negative parity and the coupling terms must be of positive parity. Also the total product in (19) must transform as the scalar representation Γ_1. Sigrist and Ueda have analyzed all possible coupling terms. For two-dimensional systems, the coupling terms can be written as in Table 6. [The extended s-wave s_* couples in the same way as s and it has been omitted]. For a vector in the x direction $\vec{n} \equiv (1, 0, 0)$, Fig.2, or in the y direction $\vec{n} \equiv (0, 1, 0)$, one obtains the important result that, apart from a uniform contribution g_1, only the $d_{x^2-y^2}$ representation is affected by the boundary in the amount g_3. For $g_3 > 0$ the surface produces an enhancement of d-wave pairing at the boundary. This result does not depend on details of how the boundary interacts, but only on its symmetry properties. It makes the suggestion of Müller [19] very plausible and makes point group invariant systems very similar to rotational invariant systems [20]. Returning to the integral equation of Ambegaokar et al., one can make reasonable models of how the order parameter changes with x. One such a model is that in which

$$\eta_{\Gamma_3} \propto \eta_0 \left[1 - \tanh \frac{x - x_0}{\xi}\right] \tag{20}$$

where ξ is of the order of the correlation length, and x_0 is the location of the boundary.

Table IV:1:6. Invariant couplings with a surface with normal $\vec{n} \equiv (n_x, n_y)$ in two-dimensions.

Γ_1	$g_1\left(n_x^2 + n_y^2\right)$	s
Γ_3	$g_1\left(n_x^2 + n_y^2\right) + g_3\left(n_x^2 - n_y^2\right)^2$	$d_{x^2-y^2}$
Γ_4	$g_1\left(n_x^2 + n_y^2\right) + g_3 n_x^2 n_y^2$	d_{xy}

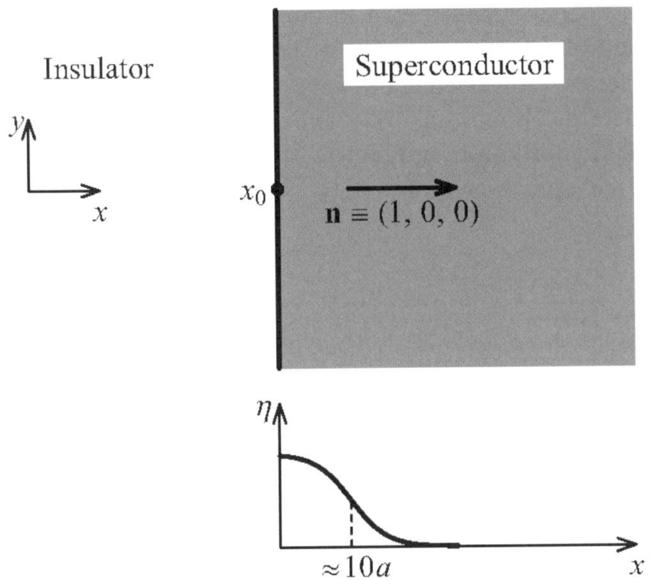

Figure IV:1:2. Geometry of the insulator-superconductor boundary discussed in the text, together with a model of the order parameter, Eq.(20)

6. BEYOND MEAN FIELD

A global phenomenological description of high-T_c superconductors is made difficult by the possibility of several almost degenerate phases and by boundary effects arising from the presence of anisotropic gap. The treatment of the same problem in atomic nuclei suggests two methods for dealing with this problem: the method of interacting bosons introduced by Arima and the author [18] and the method based on the Ginsburg-Landau formulation reported by Sigrist and Ueda [12].

In the method of interacting bosons, one introduces boson creation, $\Gamma_i^\dagger$, and annihilation, Γ_i, operators satisfying boson commutation relations

$$\left[\Gamma_i,\Gamma_j^\dagger\right]=\delta_{ij}, \quad \left[\Gamma_i,\Gamma_j\right]=\left[\Gamma_i^\dagger,\Gamma_j^\dagger\right]=0. \tag{21}$$

These boson operators are related to those introduced in [20] as shown in Table 7. They transform under D_{4h} as the representation indicated by the appropriate letter. [An extended s-wave Γ_{1*} can also be introduced.] One then constructs the Hamiltonian by expanding it into bilinear products of creation and annihilation operators, with the constraint that H must transform as the representation Γ_1 of D_{4h}

$$\begin{aligned} H = {} & \varepsilon_1 \Gamma_1^\dagger \Gamma_1 + \varepsilon_3 \Gamma_3^\dagger \Gamma_3 + \varepsilon_4 \Gamma_4^\dagger \Gamma_4 \\ & + u_{1111} \Gamma_1^\dagger \Gamma_1^\dagger \Gamma_1 \Gamma_1 + u_{3333} \Gamma_3^\dagger \Gamma_3^\dagger \Gamma_3 \Gamma_3 + u_{4444} \Gamma_4^\dagger \Gamma_4^\dagger \Gamma_4 \Gamma_4 \\ & + u_{1133} \left(\Gamma_1^\dagger \Gamma_1^\dagger \Gamma_3 \Gamma_3 + \Gamma_3^\dagger \Gamma_3^\dagger \Gamma_1 \Gamma_1 \right) + u_{1144} \left(\Gamma_1^\dagger \Gamma_1^\dagger \Gamma_4 \Gamma_4 + \Gamma_4^\dagger \Gamma_4^\dagger \Gamma_1 \Gamma_1 \right) \\ & + u_{3344} \left(\Gamma_3^\dagger \Gamma_3^\dagger \Gamma_4 \Gamma_4 + \Gamma_4^\dagger \Gamma_4^\dagger \Gamma_3 \Gamma_3 \right) \end{aligned} \tag{22}$$

This is done by using the multiplication rules for D_{4h}, for example,

$$\Gamma_3 \otimes \Gamma_3 = \Gamma_1 . \tag{23}$$

The Hamiltonian is then diagonalized in the basis

$$B : \Pi_i \left(\Gamma_i^\dagger \right)^{N_i} |0\rangle \tag{24}$$

with $N = N_1 + N_3 + N_4$. This basis is spanned by the representations of the group $U(3)$. It has been shown recently [21], that even if $N \sim 20$ the solutions are very close to the limit $N \to \infty$. Within this approach, space anisotropy can be taken into account by making $\varepsilon = \varepsilon(\vec{r})$ and $u = u(\vec{r})$. In addition one can easily discuss the case in which the coefficients depend on an external parameter, such as doping, $\varepsilon = \varepsilon(c)$ and $u = u(c)$. If both effects are present

$$\begin{aligned} \varepsilon &= \varepsilon(\vec{r}, c), \\ u &= u(\vec{r}, c). \end{aligned} \tag{25}$$

The values of ε and u will depend on the material. Diagonalization of the Hamiltonian will produce wave functions of the type

$$|\Psi\rangle = \sum_{N_1,N_3,N_4} \alpha_{N_1,N_3,N_4} \left(\Gamma_1^\dagger\right)^{N_1} \left(\Gamma_3^\dagger\right)^{N_3} \left(\Gamma_4^\dagger\right)^{N_4} |0\rangle \tag{26}$$

with

$$\alpha_i = \alpha_i(\vec{r}, c), \tag{27}$$

as advocated by Müller. With these wave functions one can evaluate the expectation values of the number operators

$$v_i = \langle\Psi|\frac{N_i}{N}|\Psi\rangle \tag{28}$$

Table IV:1:7. Boson operators in the method of interacting bosons

$s-boson$	$s^\dagger$	$\Gamma_1^\dagger$
$d_{x^2-y^2}-boson$	$\left(d_-^\dagger + d_+^\dagger\right) = d_3^\dagger$	$\Gamma_3^\dagger$
$d_{xy}-boson$	$\left(d_-^\dagger - d_+^\dagger\right) = d_4^\dagger$	$\Gamma_4^\dagger$

which give the composition of the superconducting state in terms of the representations Γ_1, Γ_3, Γ_4.

A major question however is how to construct the phase diagram of mixed systems. In order to do that, one needs to enlarge the boson space, by introducing an auxiliary boson, Γ_0, that transforms as the representation Γ_1, and represents pairs in the normal phase. The Hamiltonian becomes

$$H = \sum_{i=0,1,3,4} \varepsilon_i \Gamma_i^\dagger \Gamma_i^\dagger + \sum_{i<j=0,1,3,4} u_{iijj} \left(\Gamma_i^\dagger \Gamma_i^\dagger \Gamma_j \Gamma_j + \Gamma_j^\dagger \Gamma_j^\dagger \Gamma_i \Gamma_i\right) \tag{29}$$

and is diagonalized in the basis $\mathcal{B}$ with $N = N_0 + N_1 + N_3 + N_4$. The study of phase transitions for this system has been extensively investigated. One introduces coherent states [22]

$$|N, \eta_i\rangle = \left(\Gamma_0^\dagger + \sum_{i=1,3,4} \eta_i \Gamma_i^\dagger\right)^N |0\rangle \tag{30}$$

that depend on the (generally complex) order parameter η_i. By evaluating the expectation value of the Hamiltonian in the state (30) one obtains the energy functional

$$F(\eta_i) = \langle N, \eta_i \mid H \mid N, \eta_i \rangle. \tag{31}$$

Minimization with respect to η_i gives the equilibrium values. A study of the equilibrium values and their derivatives with respect to the coupling constants ε, *u* gives then the phase diagram [18]. If temperature dependence needs to be studied, it can be done by making the coupling constants ε, *u* temperature dependent.

An alternative method is to start directly from the Ginzburg-Landau approach [12]. In this method, one introduces (in general complex) order parameters η_i, as in Table 8, and expands the free energy in powers of the order parameters. For real order parameters and up to quartic terms one obtains

$$F(\eta_i) = \sum_{i=1,3,4} A_i(T)\eta_i^2 + \sum_{i<j=1,3,4} \beta_{ij}\eta_i^2\eta_j^2 \tag{32}$$

Table IV:1:8. Order parameters in the Ginzburg-Landau approach

s	Γ_1	$\lvert\eta_1\rvert e^{i\varphi 1}$
$d_{x^2-y^2}$	Γ_3	$\lvert\eta_3\rvert e^{i\varphi 3}$
d_{xy}	Γ_4	$\lvert\eta_4\rvert e^{i\varphi 4}$

where the notation of Sigrist and Ueda has been used. The coefficient of the second order term is temperature dependent and written as

$$A_i(T) = a' \frac{T}{T - T_{c,i}} \tag{33}$$

where $T_{c,i}$ is the critical temperature for phase Γ_i. The coefficients are material dependent and they may also depend on external parameters such as doping, *c*. The Ginzburg-Landau theory is equivalent to the method of interacting bosons.

7. CONCLUSIONS

In this article, the symmetry of superconducting phases for two-dimensional systems with D_{4h} intrinsic group has been discussed. An important result has been obtained by purely symmetry arguments namely that $d_{x^2-y^2}$ symmetry is enhanced at the surface due to boundary effects, making the recent suggestion of Müller very plausible. Finally a novel method has been introduced that allows a detailed phenomenological study of the phase structure of cuprate materials, including

(i) Anisotropy in momentum space, $\vec{k}$
(ii) Anisotropy in coordinate space, $\vec{r}$ and
(iii) Mixing of two or more almost degenerate superconducting phases.

By applying these methods to the analysis of experiments, it should be possible to understand whether or not:

(i) cuprate superconductors are anisotropic in momentum space and what is their symmetry type (s- versus d-wave)
(ii) cuprate superconductors are anisotropic in coordinate space (surface versus bulk)
(iii) different superconducting phases are mixed.

The remaining important aspect is how to derive these properties from a microscopic theory. In this respect, particularly interesting is the interacting boson-fermion model described at this workshop by Micnas [23]. This model is an extension of the method discussed above to mixed systems of bosons and fermions [24]. A symmetry analysis of this system will be presented elsewhere.

ACKNOWLEDGEMENTS

I wish to thank Alex Müller for having set me on the study of symmetry properties of high-T_c superconductors and for his continuous interest in the developments described here. This work was performed in part under D.O.E. Contract No. DE-FG-02-91 ER40608.

REFERENCES

1. J.G. Bednorz and K.A. Müller, Z. Phys. B**64**, 189 (1986).
2. J. Bardeen, L.N. Cooper and J.R. Schrieffer, Phys. Rev. **108**, 1175 (1957).
3. N.N. Bololiubov, Sov. Phys. JEPT **7**, 41 (1958).
4. J.G. Valatin, Phys. Rev. **122**, 1012 (1961).
5. P.W. Anderson, Phys. Rev. **112**, 1900 (1958).
6. R.A. Bohr, B.R. Mottelson and D. Pines, Phys. Rev. **110**, 936 (1958).
7. P.W. Anderson and P. Morel, Phys. Rev. **123**, 1911 (1961).
8. R. Balian and N.R. Werthamer, Phys. Rev. **131**, 1553 (1963).
9. A.J. Leggett, Rev. Mod. Phys. **47**, 331 (1975).
10. A. Arima and F. Iachello, Phys. Rev. Lett. **35**, 1069 (1975).
11. F. Steglich, J. Aarts, C.D. Bredl, W. Lieke, D. Meschede, W. Franz, and H. Schafer, Phys. Rev. Lett. **43**, 1892 (1979).
12. M. Sigrist and K. Ueda, Rev. Mod. Phys. **63**, 239 (1991).
13. M. Hamermesh, Group Theory, Addison-Wesley, Reading, Mass. (1962), p. 342.
14. G.F. Koster, J.O. Dimmock, R.G. Wheeler, and H. Statz, Properties of the Thirty-two point groups, M.I.T. Press, Cambridge (1963).
15. Z.-X. Shen, D.S. Dessau, B.O. Welles, D.M. King, W.E. Spicer, A.J. Arko, D. Marshall, L.W. Lombardo, A. Kapitulnik, P. Dickinson, S. Doniach, J. DiCarlo, A.G. Loeser, and C.H. Park, Phys. Rev. Lett. **70**, 1553 (1993).
16. H. Ding, T. Yokoya, J.C. Campuzano, T. Takahashi, M. Renderia, M.R. Norman, T. Mochiku, K. Kadowaki and J. Giapintzakis, Nature **382**, 51 (1996).
17. V. Ambegaokar, P.G. deGennes, and D. Rainer, Phys. Rev. A **9**, 2676 (1974).
18. For a review, see, F. Iachello and A. Arima, The Interacting Boson Model, Cambridge University Press, Cambridge, England (1987).
19. K. A. Müller, Phil. Mag. Lett. **82**, 279 (2002).
20. F. Iachello, Phil. Mag. Lett. **82**, 289 (2002).
21. F. Iachello and N.V. Zamfir, Phys. Rev. Lett **92**, 212501(2004).
22. A.E.L. Dieperink, O. Scholten and F. Iachello, Phys. Rev. Lett. **44**, 1747 (1980).
23. R. Micnas, S. Robaszkiewicz, and A. Bussmann-Holder, Physica C **387**, 58 (2003).
24. For a review, see, F. Iachello and P. van Isacker, The Interacting Boson-Fermion Model, Cambridge University Press, Cambridge, England (1991)

IV.2

EVIDENCE FOR *d*-WAVE ORDER PARAMETER SYMMETRY IN Bi-2212 FROM EXPERIMENTS ON INTERLAYER TUNNELING

Yu I. Latyshev
Inst. of Radio-Engineering and Electronics, Russian Acad. of Sci., Mokhovaya 11-7, Moscow 101999

Abstract: We consider three group experiments on interlayer tunneling on Bi-2212 mesa-type structures, fabricated by focused ion beam (FIB) technique from Bi-2212 single crystal whiskers, pointing out to the *d*-wave type of the order parameter (OP) in this compound. We specify the experiments on low temperature interlayer quasiparticle conductivity and magneto-conductivity on small Bi-2212 mesas and the experiments on Josephson flux-flow dissipation on long Bi-2212 mesas. All the results are shown to be consistent with a *d*-wave Fermi-liquid model with a significant contribution of coherent interlayer tunneling.

Key words: $Bi_2Sr_2CaCu_2O_{8+x}$ *c*-axis Transport and Magneto-transport, Single Crystal Whiskers, *d*-wave, Order Parameter Symmetry, Josephson Flux-flow Resistivity.

1. INTRODUCTION

More than 15 years studies of high temperature superconductivity in cuprates accumulated many evidences of the *d*-wave type symmetry of the superconducting order parameter (OP) in these materials. The most strong ones has been found from the ARPES experiments [1], quantum interference

A. Bianconi (ed.), Symmetry and Heterogeneity in High Temperature Superconductors, 181–197.

on tricrystal boundary [2,3], experiments on YBCO/Nb zigzag junctions [4] etc. Here we consider several experiments on interlayer tunneling on Bi-2212 mesa-type structures that also points out to the *d*-wave symmetry of the OP. The advantage and the complement of this method to other techniques is that that is essentially bulk method. The experiments considered below have been carried out on the *c*-axis structures fabricated from single crystal whiskers, the very perfect single crystal objects [5,6].

The interlayer tunneling technique [7,8] is based on the layered crystalline structure of the highly anisotropic high temperature superconductors as Bi- and Ta-based materials. Microscopically this structure represents a stack of elementary superconducting layers separated by elementary oxide layers that forms a series of elementary tunneling junctions. To study the interlayer Josephson tunneling correctly one needs to have a stack structure of the micron scale lateral size, L, $L < \lambda_J = s\ \lambda_c/\lambda_{ab}$,, where s is the spacing between elementary superconducting layers (s = 15.6 Å for Bi-2212) and $\lambda_{ab,c}$ is the anisotropic London penetration depth ($\lambda_c/\lambda_{ab} \approx$ 500 for slightly overdoped Bi-2212), that contains also a small enough number of elementary junctions, N, typically N being of few tens. In the opposite limit of long junctions $L >> \lambda_J$ the interlayer Josephson tunneling is essentially affected by the presence of Josephson vortices.

2. FABRICATION OF THE c-AXIS STRUCTURES

The Bi-2212 single crystal whiskers have been grown by impurity-free method [6]. Thin whiskers have been characterized as a very perfect single crystal objects. They grow along the *a*-axis free of any crucibles or substrates and can be entirely free of macroscopic defects and dislocations. They can reach in a length the size up to 10 mm having, however, the typical size below 1mm. In the *bc*-plane they have rectangular cross-section with typical sizes L_b=1-30 μm, L_c= 0.1-3 μm.

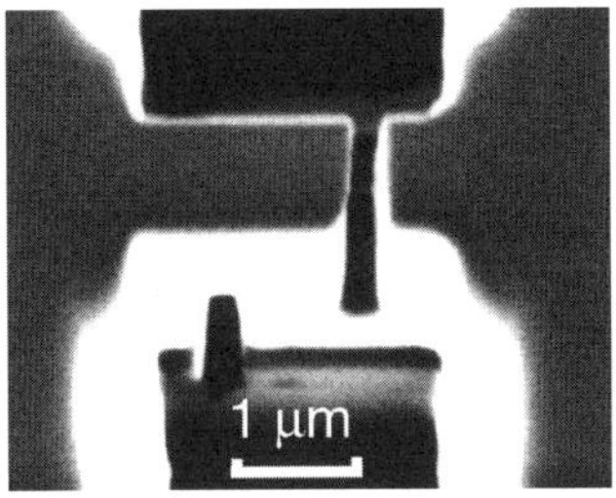

Figure IV:2:1. SEM picture of the typical Bi-2212 stacked structure, z-direction corresponds to the c-axis

For the fabrication of stacked junctions we used focused ion beam technique. This technique has been developed for fabrication of both, the short [9] down to submicron scale [10] and the long junctions with a length of several tens microns [17]. For fabrication we used conventional FIB machine of Seiko Instr. Corp., SMI 9800 (SP) with Ga+-ion beam. The four leads were attached outside the junction area. The contact Ag pads were ablated and annealed before the FIB processing to avoid diffusion of Ga-ions into the junction body. The example of a short stack fabricated by FIB technique is shown in Fig. 1. Typically we had slightly overdoped stacked $Bi_2Sr_2CaCu_2O_{8+\delta}$ structures with $\delta \approx 0.25$. They have $T_c = 77K$, $\rho_c(300K) = 10\text{-}12$ Ohm cm, $J_c(4.2K) \approx 1$ kA/cm^2.

3. THE c-AXIS TRANSPORT ON SHORT STACKS

Here we consider results on short stacks with typical lateral size 1 μm x 1 μm and containing $N = 30\text{-}70$ elementary junctions. The high quality of the mesas has been approved by the Fraunhofer patterns of critical current I_c on parallel magnetic field with periodicity of one flux per elementary junction [11, 12]. Fig 2 shows the I-V characteristics of the short stacked junctions in large and small voltage scales.

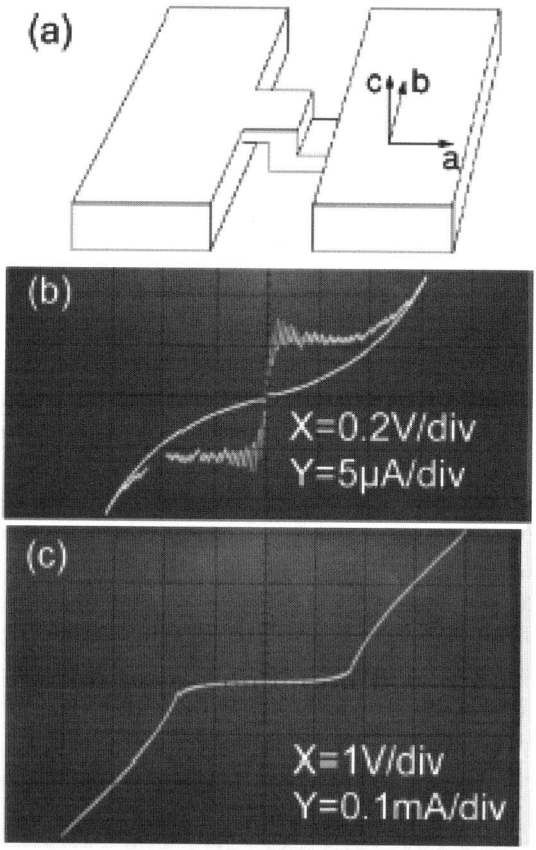

Figure IV:2:2. Schematic view of the junction (a); and the I-V characteristics of the Bi-2212 stacks in (b) enlarged scale for sample #3 [14] and (c) extended scale for sample #4 [12]. T = 4.2 K

The superconducting gap voltage, V_g, (Fig.2c) was determined as the voltage of the maximum of the *dI/dV(V)*. The gap of intrinsic junction $2\Delta_0 \approx eV_g/N$ reaches value as high as 50 meV [10]. The multibranched structure (Fig.2b) corresponds to subsequent transition of the intrinsic junctions into the resistive state for increasing voltage [7]. At voltages $V>V_g$ all junctions are resistive. Therefore in downsweep of voltage, starting from $V>Vg$, the *I-V* curve is observed in the all junction resistive state. Here only quasiparticles contribute to the *c*-axis transport. Corresponding quasiparticle conductivity, σ_q, thus can be defined directly from that part of the *I-V* curve.

The Ohmic resistance, R_n, at $V > V_g$ is also well defined from the *I-V* characteristics (Fig. 2c). This resistance is nearly temperature independent and corresponds to the conductivity $\sigma_n(V>V_g) \approx 80$ (kOhm cm)$^{-1}$ for energies above superconducting gap. The critical current I_c was determined from the *I-V* characteristics as the current of switching from the superconducting to the resistive state, averaged over the stack. The variation of the critical current along the stack is not large (usually within 15%), indicating a good uniformity of our structures. The *c*-axis critical current density was typically ~ 1kA/cm^2 [12, 13].

We found out that interlayer tunneling *I-V* characteristics at low temperatures essentially differ from those of conventional Josephson junctions between s-wave superconductors. We specify [14]: 1) $J_c(0)$ is strongly reduced to compare with the value expected from Ambegaokar-Baratoff (A-B) relation, $J_c^{AB}(0) = \pi\sigma_n\Delta_0 / 2es$; 2) quadratic and scaling behaviour of $\sigma_q(V,T)$ on V and T (Figs. 3,4):

$$\sigma_q(V,T) = \sigma_q(0,0)\begin{cases}(1+bV^2)\\(1+cT^2)\end{cases} \tag{1}$$

with $b = 0.014 \pm 0.03$ mV^{-2}, $c = (6 \pm 2) \times 10^{-4}$ K^{-2} and $c/b \sim 4$; 3) nonzero and universal value of $\sigma_q(0,0) \approx 2.5$ (kOhm cm)$^{-1}$.

We found empirically the modified relation of the A-B type between J_c and σ_q:

$$J_c \approx \pi\sigma_q(0,0)\Delta_0 / 2es \tag{2}$$

It was shown then that all these observed features can be described self-consistently by Fermi-liquid model for quasiparticles in clean *d*-wave superconductor with resonant intralayer scattering [14]. The superconducting gap is expressed as $\Delta(\varphi) = \Delta_0 \cos 2\varphi$, where φ is the angle of the momentum on the two-dimensional cylindrical Fermi-surface.

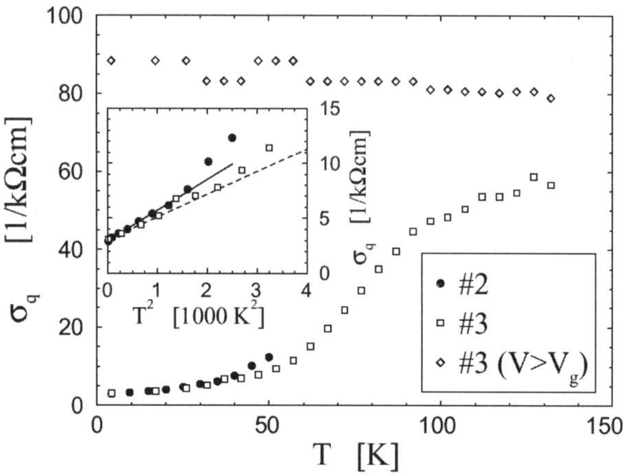

Figure IV:2:3 Quasiparticle dynamic conductivity σ_c vs T for voltages $v > V_g/N \approx 2\Delta_0/e$ and $v = V/N \rightarrow 0$ as extended from the *I-V* characteristics of samples #2 and #3 [14]. The insert shows σ_c vs T^2 at $v \rightarrow 0$. Lines are fit for $T^2 < 1000\ \mathrm{K}^2$

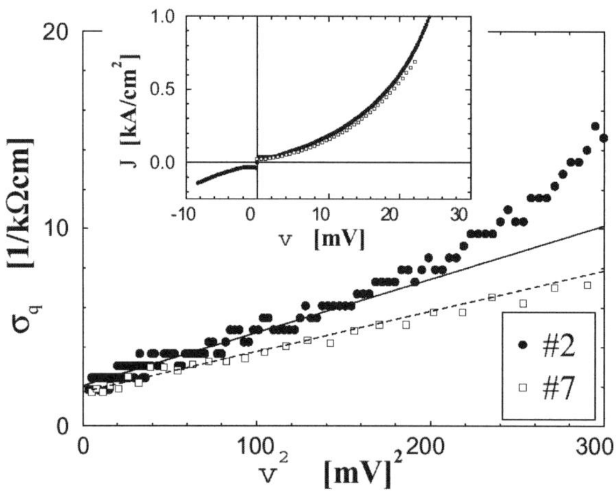

Figure IV:2:4. The quasipartical differential conductivity $v^2 = V^2 / N^2$ at T = 4.2 K as extracted from the *I-V* characteristics of sample #2 and #7 (Fig. 4c in Ref. [8]). Lines are fits for $v < 10$ mV. Insert: Corresponding *J-v* curves

Impurity scattering leads to the formation of the gapless state at some sectors φ_0 near the node directions φ_g at angles $\varphi_g \pm \varphi_0/2$, with $\varphi_0 \sim \gamma/\Delta_0$, where γ is the impurity bandwidth of quasiparticles. That results in a nonzero density of states at zero energy, $N(0)$, where $N(0)$ is the 2D density of states per spin at the Fermi level, and leads to a universal quasiparticle interlayer conductivity $\sigma_q(0,0)$:

$$\sigma_q(0,0) \propto \frac{N(0)\varphi_0}{\gamma} = \frac{N(0)}{\Delta_0} \,. \tag{3}$$

As that follows from Eq. (3), $\sigma_q(0,0)$ is independent on impurity scattering rate γ. The expression (3) is valid for the coherent interlayer tunneling that conserves the in-plane momentum in tunneling process. In more general case one can introduce coherence factor characterizing a weight of coherent tunneling as a while the weight of incoherent one as (1- a). Then one can get the following expressions for $J_c(0)$, $\sigma_q(0,0)$, b and c [14]:

$$\sigma_q(0,0) = \frac{2e^2 t_\perp^2 N(0) s}{\pi\hbar\Delta_0}[a + (1-a)\frac{\gamma}{\varepsilon_F}]; \quad J_c(0) = \frac{e t_\perp^2 N(0)}{\hbar}[a + (1-a)\frac{\Delta_0}{\varepsilon_F}] \tag{4}$$

This theory also gives the scaling behaviour of $\sigma_q(V,T)$ of the type of Eq. (1):

$$b = \frac{1}{8\gamma^2}[1 + \frac{(1-a)}{a}\frac{\gamma}{\varepsilon_F}]; \; c = \frac{\pi^2}{18\gamma^2}[1 + \frac{(1-a)\gamma}{\varepsilon_F}], \tag{5}$$

where $t_\perp$ is a tunneling matrix element [14]. One can see that the ratio $J_c(0)/\sigma_q(0,0)$ from Eq. (4) turns to the experimentally found expression (2) only for significant contribution of coherent tunneling, $a >> \max\{\Delta_0/\varepsilon_F, \gamma/\varepsilon_F\}$. From experimental value for c we can estimate γ using Eq. (5) as to be $\gamma \approx$ 3mV. Then for $\Delta_0/\varepsilon_F \approx 0.1$ we can get estimation for a, $a >> 0.1$. In that case $c/b = 4\pi^2/9 \approx 4.4$ in a good agreement with experiment. Also, for $\sigma_q(0,0)$ we can get from Eq. (4), using J_c = 0.8 kA/cm^2, Δ_0 =30 meV, s =1.5 10^{-7} cm, the value 2.6 (kOhm cm)$^{-1}$ which is very close to the experiment.

Thus we can infer that experimental low *V, T* interlayer tunneling data are consistent with Fermi-liquid picture taking into account *d*-wave symmetry and essentially coherent interlayer tunneling.

4. THE C-AXIS MAGNETO-CONDUCTANCE IN BI 2212

In this part we consider the *c*-axis magneto-transport on single crystals and small mesas in high magnetic field perpendicular to the layers. In both cases we access qasiparticle current by suppressing Josephson interlayer current in one of two ways: (a) by magnetic field of 60T in a case of single crystals, (b) by current that driving a small mesa in the all junction resistive state, as that was discussed above.

The *c*-axis magneto-resistance of Bi-2212 single crystals at different temperatures is shown in Fig. 5.

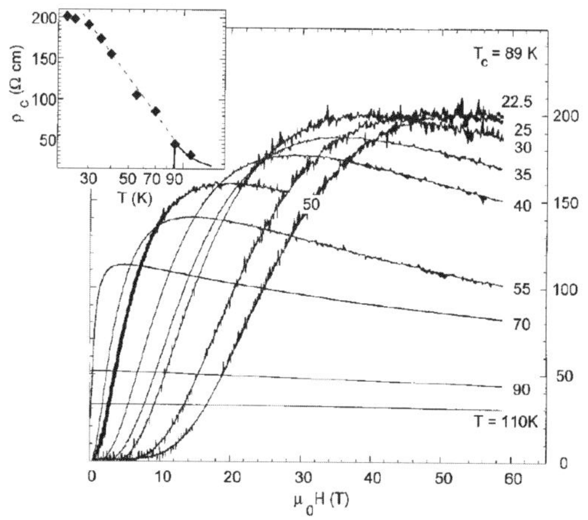

Figure IV:2:5. The out-of-plane resistivity ρ_c vs magnetic field of a slightly overdoped Bi-2212 crystal in fields up to 60 T at different temperat ures [15]. $J = 0.05$ A/cm^2. Insert: ρ_c vs T at 55T (full rhombus) and 0 T (line)

The characteristic feature is that magneto-resistance achieves a maximum at some field and then drops down demonstrating negative magneto-resistance.

Fig.6 shows the data of magneto-conductivity versus magnetic field partly including the data of Fig.5. This graph shows two contributions to magneto-conductivity $\sigma_c(H,T)$[15]: (1) contribution of Josephson tunneling σ_J which drops rapidly with field as $\sigma_J = \alpha H^{\nu}$, where the power index ν varies from 1.5 to 3.5 with temperature variation from 70K to 22.5K ; (2) contribution of quasiparticles $\sigma_q(H,T) = \sigma(0,T)\,[\eta + \beta H]$, where $\sigma_q(0,T) = \sigma_q(0,0)(1 + cT^2)$ with $\sigma_q(0,0)$ =2.5 (kOhm cm)$^{-1}$ as that was discussed before [14]. Both contributions plotted for T = 55K at Fig. 6 quite well describe experimental dependence of $\sigma_q(H,T)$ [15]. As shown at high enough field $\sigma_J(H)$ becomes negligibly small and the linear growth of $\sigma_q(H)$ or negative magneto-resistance is completely defined by quasiparticle contribution. That conclusion has been confirmed by studies of magneto-resistance on mesas in all junctions resistive state, Fig. 7. Thus experiment shows a linear growth of quasiparticle conductivity with field $\sigma_q(H,T) - \sigma_q(0,T) \propto \sigma_q(0,0)\,H$. This unusual behaviour has been explained in a *d*-wave Fermi-liquid model [16]. It was shown that at high field limit the variation of quasiparticle conductivity, $\Delta\sigma$, is expressed as follows [16], $\Delta\sigma_q(H) \sim \sigma_q(0,0)\,H/H_{\Delta}$, where $H_{\Delta} = \Phi_0\Delta_0 / \hbar^2 v_F^2 \approx$ 40-80 T. This dependence is consistent with experiment [15] (see Figs. 6,7)

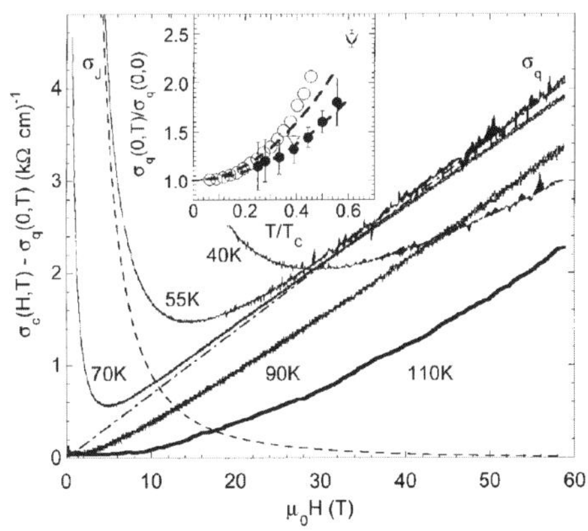

Figure IV:2:6. $\sigma_c(H)$ below and above T_c = 89 K. A fit at 55 K to a superposition of Cooper pair (dashed line) and quasiparticle (dash-dotted line) contributions to magneto-conductivity is indicated (see text). Insert: Zero field σ_c vs T/T_c extracted from $\sigma_c(H,T)$ for a Bi-2212 crystal for J = 0.05 A/cm^2 (full circles) and J = 0.1 A/cm^2 (triangles), and obtained from the *I-V* curves for a mesa (empty circles). Both fit a T^2 dependence (dashed line) up to $T = \gamma$

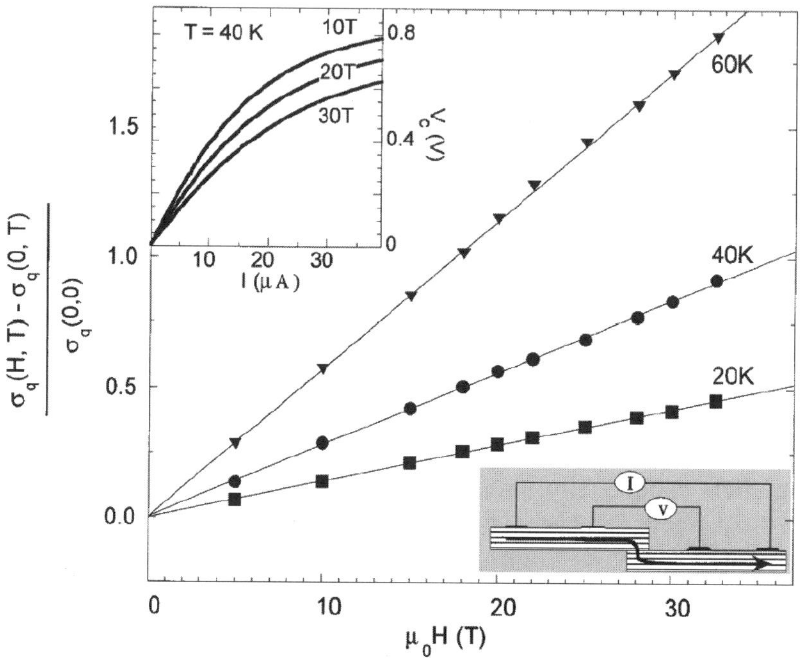

Figure IV:2:7. Normalized quasiparticle *c*-axis conductivity as a function of *H//c* obtained from the *I-V* curves (top inset) measured on the mesa-shaped Bi-2212 (sketched) [15]

5. JOSEPHSON FLUX-FLOW IN-PLANE DISSIPATION IN Bi-2212

In this part we describe our experiments on studies of dissipation in Josephson flux-flow regime. We show that experiments of that type can be used for studies of temperature dependence of both the in-plane and the out-of-plane components of the quasiparticle conductivity [17].

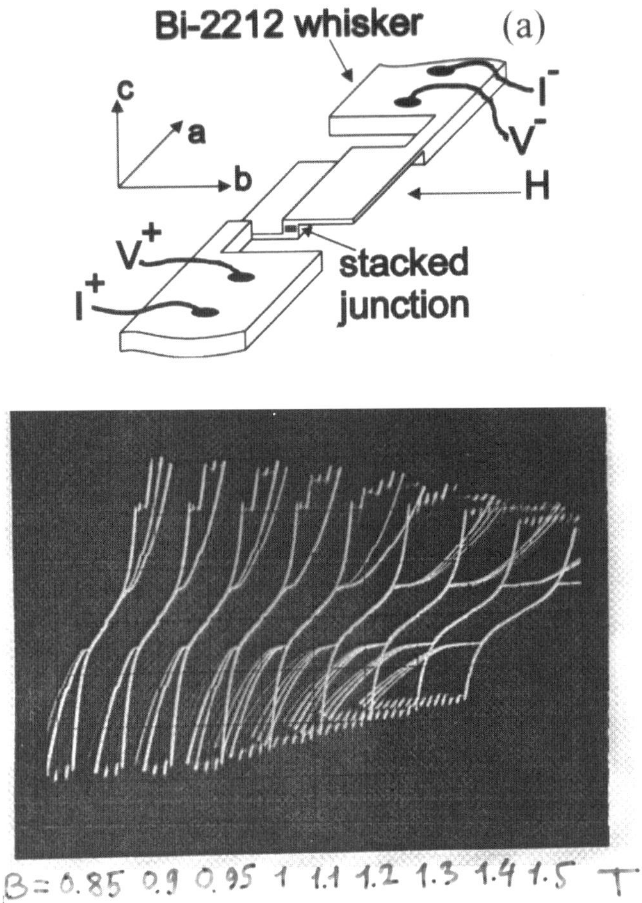

Figure IV:2:8. The set up (top insert) and the set of the *I-V* characteristics measured in the flux-flow regime with field *B* applied along the b-axis, *B* increasing from 0.85 T up to 1.5 T

The unusual temperature dependence of $\sigma_{ab}(T)$ found from our experiment [17] is consistent with the microwave data [18] and can be

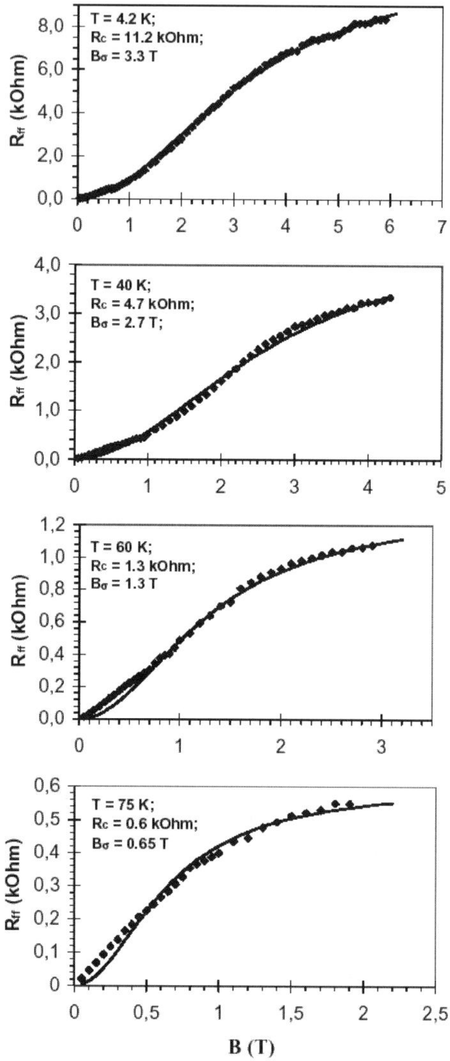

Figure IV:2:9. Magnetic field dependence of the Josephson flux-flow resistance, R_{ff}, at different temperatures with fits to Eq. (7) [17]

naturally explained by a *d*-wave Fermi-iquid model. As that is well known, magnetic field applied parallel to the layers of the layered high-T_c superconductor like Bi-2212 (below we consider a long enough mesa-structure sketched at Fig.8) can introduce the Josephson vortices (JVs) [19] with characteristic sizes λ_J along the layers and λ_{ab} across the layers. They are centered between superconducting layers and have no normal core. A concentration of Josephson vortices grows with field and the interaction between them leads to formation of triangular Josephson vortex lattice (JVL) [20]. With field growth at $H > H_0 = \Phi_0 / \lambda_J s$ non-linear cores of Josephson vortices in one line start to overlap and so called dense JVL is formed. JVs strongly interact with each other in a dense JVL. As a consequence, when a DC current is applied across the layers, the resulting Lorentz force can drive a dense JVL as a whole [21]. This regime is known as Josephson flux-flow (JFF). That is characterized by a specific JFF branch on the IV characteristic having non-linear upturn with maximum voltage being proportional to the applied magnetic field (Fig. 8) [22]. In a linear JFF regime the dissipation is proportional to the linear slope of JFF branch at V→ 0 (Fig. 8).In a general case that can be expressed as follows [23]:

$$\sigma_{JFF} E_z^2 = \sigma_c E_z^2 + \sigma_c E_{z\omega}^2 + \sigma_{ab} E_{x\omega}^2 . \qquad (6)$$

Here directions z and x correspond to the orientation of electric field across and along the layers correspondingly, $E_{x,z}$ is the DC components of electric field, index ω is related to the AC electric field components associated with the motion of JVL. In earlier calculations of dissipation in JFF regime [19, 24] only two first terms of Eq. (6) have been taken into account. However, as that has been pointed out recently [23], the third term can have the leading contribution for the highly anisotropic layered superconductors, satisfying the condition $\Gamma = (\sigma_{ab} / \sigma_c)/(\lambda_c^2 / \lambda_{ab}^2) > 1$. For materials like BSCCO at low temperatures Γ is typically about 50. In this case the JFF resistivity was predicted to have unusual quadratic dependence on parallel magnetic field B [23]:

$$\rho_{Jff} = \frac{B^2}{B^2 + B_\sigma^2}\rho_c , \quad B_\sigma = \sqrt{\frac{\sigma_{ab}}{\sigma_c}} \frac{\Phi_0}{\sqrt{2}} \frac{1}{\pi \gamma^2 s^2} \qquad (7)$$

The experiment shows very nice fit of $\rho_{JFF}(B)$ dependence to this expression [17] at wide temperature range, as shown at Fig. 9.

That allows us to extract from the JFF experiment both components of quasiparticle conductivity σ_c and σ_{ab} and their temperature dependences [17] (Fig.10). As shown, they both are consistent with those measured by other independent methods.

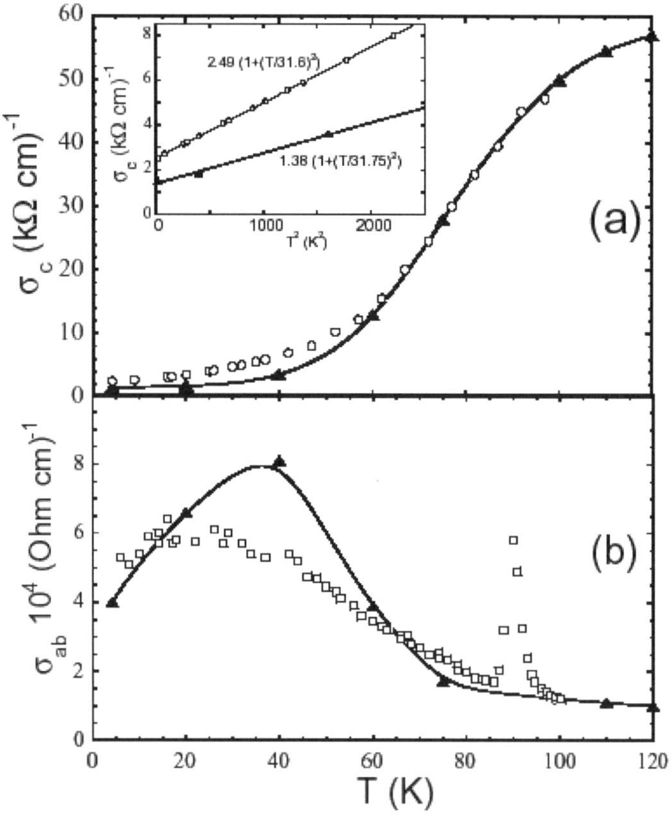

Figure IV:2:10. Solid triangles show temperature dependence of the out-of-plane quasiparticle conductivity σ_c (a) and in-plane quasiparticle conductivity σ_{ab} (b). Below T_c they are extracted from the JFF experiment on BSCCO long stack [17] and above T_c they represent the normal state conductivities of whiskers measured independently on samples from the same batch. Open circles correspond to the σ_c data from Ref.14, obtained on small mesas in zero field, open squares correspond to 14.4 GHz microwave data for σ_{ab} from Ref. 18 obtained on epitaxial films. Solid lines in both plots are just guides for the eye. Insert in (a) shows the low temperature part of $\sigma_c(T)$ plotted versus T^2

The most important result is that temperature dependence of the in-plane quasiparticle conductivity, $\sigma_{ab}(T)$, well reproduces unusual maximum at T = 20-30K. This type of $\sigma_{ab}(T)$ behaviour has been found earlier in the microwave experiments for YBCO [25] and BSCCO [18] and also is consistent with the heat transport measurements of the electronic part of the thermal conductivity [26].

Fig.10b shows a comparison of our data with the microwave results [18]. The origin of the peak of $\sigma_{ab}(T)$ has been widely discussed as a result of a *d*-wave symmetry of the OP in BSCCO and YBCO. In particular, in a *d*-wave Fermi-liquid model it was shown that at low temperatures σ_{ab} grows with temperature as [27]: $\sigma(\omega\rightarrow 0,T) = \sigma_{00}\ (1 + \gamma T^2)$, where σ_{00} is a universal in-plane conductivity introduced by Lee [28], $\sigma_{00} = n\ e^2\ /(\pi\ m_{ab}\ \Delta_0)$ with m_{ab} the effective quasiparticle in-plane mass.
Actually this growth is caused by thermally activation of nodal quasiparticles. At higher temperatures the quasiparticle relaxation time drops down [29]. The peak in a temperature dependence of σ_{ab} appears then as a result of an interplay between the temperature dependences of the relaxation rate and concentration of quasiparticles.

6. CONCLUSIONS

The experiments on interlayer tunneling on short and long Bi-2212 mesas show new evidences of the *d*-wave OP symmetry. That results in: (1) the highly reduced value of the *c*-axis critical current density in comparison with the expected value from conventional Ambegaokar-Baratoff relation, (2) finite and universal value for the *c*-axis low temperature quasiparticle conductivity and its scaling relations on the bias voltage and temperature, (3) linear growth of magneto-conductivity in high magnetic fields, (4) the unconventional temperature dependence of the in-plane conductivity having a peak at 20-30K.

AKCNOWLEDGEMENTS

We acknowledge support from the CRDF grant No RP1-12397-MO-02, grant from Russian Ministry of Science and Industry No. 40.012.1.111.46 and Jumelage project between IRE RAS and CRTBT CNRS, No. 03-02-2201.

REFERENCES

1. Z-X. Shen et al. Phys. Rev. Lett. 70, 1553 (1993); Z-X. Shen et al., Science, 267, 343 (19995).
2. C.C. Tsuei et al. Phys. Rev. Lett. 73, 593 (1994).
3. J.R. Kirtley et al. Nature, 373, 225 (1995); C.C. Tsuei and J.R. Kirtley, Rev. Mod. Phys., 72, 969 (2000).
4. H.J.H. Smilde et al. Phys. Rev. Lett., 88, 05704 (2002).
5. I. Matzubara, H. Kageyama, T.Ogura, H. Yamashita, and T. Kawai, Jpn. J. Appl. Phys. 28, L 1121 (1989).
6. Yu.I. Latyshev, I.G. Gorlova, A.M. Nikitina, V.U. Antokhina, S.G. Zybtsev, N.P. Kukhta and V.N. Timofeev, Physica C 216, 471 (1993).
7. R. Kleiner, F.Steinmeyer, G. Kunkel, P.Muller, Phys. Rev. Lett., 68, 239, (1992); R.Kleiner and P. Mueller, Phys. Rev. B 49, 1327 (1994).
8. K. Tanabe, Y. Hidaka, S. Karimoto, M. Suzuki, Phys. Rev. B 53, 9348 (1996).
9. Yu.I. Latyshev, S.-J. Kim, T. Yamashita, IEEE Trans. Appl. Supercond., 9, 4312 (1999); S.-J. Kim, Yu.I. Latyshev and T. Yamashita, Supercond. Sci. Technol. 12, 728 (1999).
10. Yu.I. Latyshev, S.-J. Kim, T. Yamashita, JETP Lett., 69, 84 (1999).
11. L.N. Bulaevskii, J.R. Clem, L.I. Glazman, Phys. Rev. B 46, 350 (1992).
12. Yu.I. Latyshev, S.-J. Kim, V.N. Pavlenko, T. Yamashita, L.N. Bulaevskii, Physica C, 362, 156 (2001).
13. K. Inomata, T. Kawae, K. Nakajima, S.-J. Kim, and T. Yamashita, Appl. Phys. Lett. 82, 769 (2003).
14. Yu.I. Latyshev, T. Yamashita, L.N. Bulaevskii, M.J. Graf, A.V. Balatsky, and M.P. Maley, Phys. Rev. Lett. 82, 5345 (1999).
15. N. Morozov, L. Krusin-Elbaum, T. Shibauchi, L.N. Bulaevskii, M.P. Maley, Yu.I.Latyshev, and T.Yamashita, Phys. Rev. Lett. 84, 1784 (2000).
16. I. Vekhter, L.N. Bulaevskii, A.E. Koshelev, and M.P. Maley, cond-mat/9910341.
17. Yu.I. Latyshev, A.E. Koshelev, L.N. Bulaevskii, Phys.Rev. B 68, 134504 (2003).
18. Shin-Fu Lee et al., Phys. Rev. Lett., 77, 735 (1996); H. Kitano et al., J. Low Temp. Phys., 117, 1241 (1999).
19. J. Clemm and M. Coffey, Phys. Rev.B 42, 6209 (1990).
20. L.N. Bulaevskii and J.R. Clem, Phys. Rev., B 44, 10234 (1991).

21. L.N. Bulaevskii, D. Domingez, M.P. Maley, A.R. Bishop, and B.I. Ivlev, Phys. Rev. B 53, 14 601 (1996).
22. J.U. Lee et al., Appl. Phys. Lett., 67, 1471 (1995); G. Hechtfischer et al. Phys. Rev. Lett., 79, 1365 (1997); Yu.I. Latyshev et al., Physica C 293, 174 (1997).
23. A.E. Koshelev, Phys. Rev., B 62, R3616 (2000).
24. S. Sakai et al. Phys. Rev., B 50, 12 905 (1994); M. Machida et al. Physica C, 330, 85 (2000).
25. D.A. Bonn et al., Phys. Rev. Lett., 68, 2390 (1992); D.A. Bonn et al., Phys. Rev. B 50, 4051 (19994); A. Hosseini et al., ibid. 60, 1349 (1999).
26. K. Krishna et al. Phys. Rev. Lett., 75, 3529 (1995); B. Zeini et al. Eur. Phys. J. B 20, 189 (2001).
27. P. J. Hirschfeld, W.O. Puttika, D.J. Scalapino Phys. Rev. B 50, 10250 (1994).
28. P.A. Lee, Phys.Rev. Lett, 71, 1887 (1993).
29. J. Corson, J. Orenstein, S. Oh, J. O'Donell, J.N. Eckstein, Phys. Rev. Lett., 85, 2569 (2000).

V

EXOTIC SUPERCONDUCTIVITY

V.1

ELECTRONIC STATE IN Co-OXIDE–SIMILAR TO CUPRATES?

S. Maekawa and W. Koshibae
Institute for Materials Research Tohoku University, Sendai 980-8577, Japan

Abstract: It is shown that the electronic structure in layered cobalt oxides with hexagonal crystal structure is described as a Kagomé lattice hidden in the CoO_2 layer which consists of stacked triangular lattices of oxygen ions and of cobalt ones. The Kagomé lattice is derived because of the degeneracy of t_{2g} orbitals. We discuss that the electronic structure causes a variety of unique properties in the cobalt oxides such as superconductivity and ferromagnetism, which are in contrast to the high-T_c cuprates.

PACS numbers: 74.25.Jb, 71.10.-w, 71.27.+a

1. INTRODUCTION

There is a growing interest in cobalt oxides (cobaltates) with layered hexagonal crystal structure. They show giant thermopower [1-4], giant Hall effect [5], ferromagnetism [6,7] depending on the material details. In particular, the superconductivity[8] has received special attention [9-21] in connection to the high-T_c superconductivity in cuprates. This is because of the following reasons: (i) The superconductivity occurs on the triangular lattice in cobaltates whereas it does on the square lattice in cuprates, (ii) ferromagnetism appears in cobaltates and antiferromagnetism does near the superconducting states in cuprates, and (iii) the active electronic states are t_{2g}

A. Bianconi (ed.), Symmetry and Heterogeneity in High Temperature Superconductors, 201–212.

with orbital degeneracy and the e_g without degeneracy in cobaltates and cuprates, respectively.

It has been pointed out that the giant thermopower in cobaltates is caused by the degeracy of the t_{2g} orbitals of Co ions [22, 23]. Here, we show that the orbital degeneracy brings about a Kagomé lattice electronic structure hidden in the CoO_2 triangular crystal lattice [15]. This is because the electron hopping occurs between Co ions via neighboriong oxygens by exchanging the orbitals in the triangular lattice. In Sec. 2, the electronic structure is examined. The unique physical properties of cobaltates are discussed in Sec.3 based on the electronic structure.

2. KAGOMÉ IN TRIANGULAR LATTICE

The CoO_2 layer is formed by the edge-shared CoO_6 octahedra which are compressed along c-axis (Fig. 1). The rhombohedral distortion of the CoO_6 octahedra is measured by the deviation of the O-Co-O bond angle from 90°, 95°~ 99° [2, 24-27]. The distortion leads to the crystal-field splitting in t_{2g} states of $3d$ electrons as shown in Fig. 1(c).

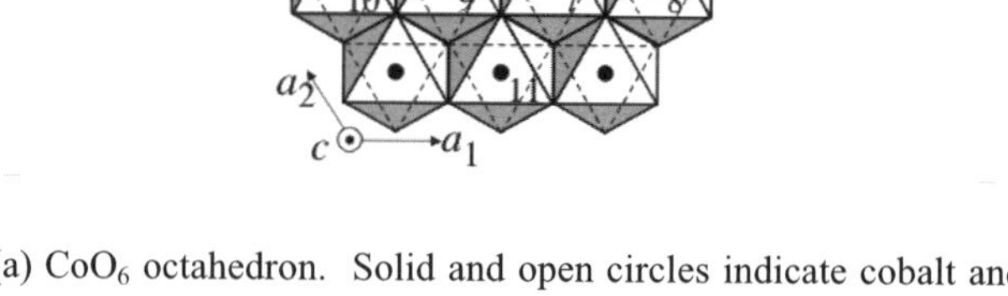

Figure V:1:1. (a) CoO_6 octahedron. Solid and open circles indicate cobalt and oxygen ions, respectively. (b) CoO_2 layer. c and a_1 axes are along (1,1,1) and (−1,1,0) directions in xyz coordinate system shown in (a). The numbers (0 ~12) on solid circles are the labels of Co sites. (c) The crystal-field splitting of the distorted CoO_6 octahedron. e'_g is used to distinguish from the e_g (x^2-y^2 and $3z^2$-r^2) states

The wave functions are expressed as

$$(|xy> + |yz> + |zx>)/\sqrt{3} \tag{1}$$

for the a_{1g} state and

$$(|xy> + e^{\pm i\frac{2\pi}{3}}|yz> + e^{\pm i\frac{4\pi}{3}}|zx>)/\sqrt{3}$$

for the doubly degenerate states where $|xy>$, $|yz>$ and $|zx>$ denote the wave functions of the t_{2g} states. The a_{1g} state extends to the c-axis whereas the e'_g states spread over the plane perpendicular to the c-axis. Since the apex oxygens approach the plane in the distorted CoO_6 octahedra, the a_{1g} state is stabilized [28] for an electron.

The band calculation [24] in $Na_{0.5}CoO_2$ has shown that the energy splitting between a_{1g} and e'_{1g} states at the Γ point is ~1.6eV which is the total band width of the t_{2g} manifold and the a_{1g} state is higher than the e'_g states.

This fact shows that the energy splitting does not originate in the crystal field due to the distortion but is determined by the kinetic energy of electrons.

Let us consider the hopping-matrix-elements between neighboring $3d$ orbitals of cobalt ions neglecting the rhombohedral distortion. There are two mechanisms for the hopping of an electron: one is the hopping integral between adjacent $3d$ orbitals, and another is owing to the hopping between a $3d$ orbital of a cobalt ion and a $2p$ one of an oxygen ion. First, let us consider the latter mechanism, i.e., the hopping of a $3d$ electron through the $2p$ orbital on the neighboring oxygen. The CoO_2 layer in the hexagonal structure is expressed as a triangular lattice of cobalt ions sandwiched by those of oxygen ions, i.e., both the upper and lower layers of oxygens form triangular lattices. The lower layer is drawn by broken lines in Fig. 1(b). In the following, the t_{2g} orbitals on the i-th cobalt ion are expressed as $|xy,i>$, $|yz,i>$ and $|zx,i>$, respectively. The CoO_6 octahedra share edges each other, so that Co-O-Co bond angle is $\pi/2$. The state $|xy,0>$ has a hopping matrix element with $|zx,7>$ through the $2p_x$ orbital of the oxygen ion which exists in the upper layer and shares CoO_6 octahedra involving cobalt ions 0 and 7, respectively (Fig. 2).

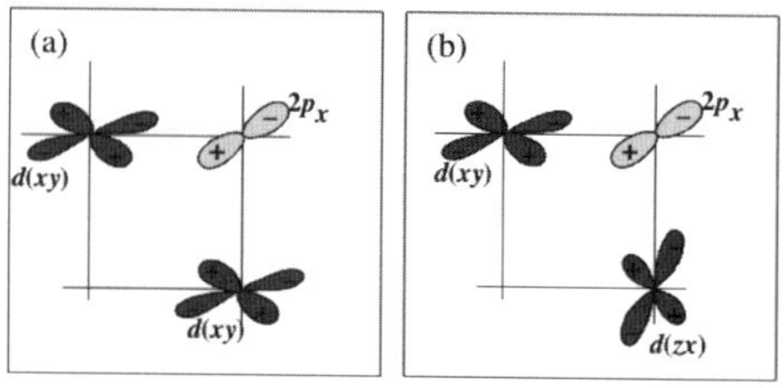

Figure V:1:2. Hopping-matrix between neighboring Co ions. (a) There is no hopping matrix between |xy,0> and |xy,7>. (b) There is the hopping matrix between |xy,0> and |zx,7>

The hopping matrix element (t) is expressed as $t \sim t_{pd}^2/\Delta(>0)$, where Δ is the energy level of the $2p_x$ orbital measured from that of t_{2g} states and t_{pd} is the hopping integral between the $2p_x$ and |xy,0> (or |zx,7>) orbitals. There also exists a hopping matrix element between |zx,0> and |xy,7> which is due to the $2p_x$ orbital of an oxygen ion on the lower layer. On the other hand, there is no hopping matrix element between |xy,0> and |xy,7> because of the symmetry. In the same way, we find the hopping matrix elements between the following pairs of orbitals: (|xy, 0>,|zx, 2>), (|zx, 0>,| xy, 2>), (| xy, 0>,|yz, 4>), and (|yz, 0>,| xy, 4>). As a result, the hopping matrices of a 3d electron in $\vec{a}_1$, $\vec{a}_2$ and $\vec{a}_1+\vec{a}_2$ directions where $\vec{a}_1$ and $\vec{a}_2$ are the elementary translation vectors along a_1 and a_2 axes, are expressed as

$$\begin{array}{c} \\ xy \\ yz \\ zx \end{array} \begin{array}{c} xy\ yz\ zx \\ \begin{pmatrix} 0 & 0 & 0 \\ 0 & 0 & t \\ 0 & t & 0 \end{pmatrix} \end{array}, \begin{array}{c} xy\ yz\ zx \\ \begin{pmatrix} 0 & 0 & t \\ 0 & 0 & 0 \\ t & 0 & 0 \end{pmatrix} \end{array} \text{ and } \begin{array}{c} xy\ yz\ zx \\ \begin{pmatrix} 0 & t & 0 \\ t & 0 & 0 \\ 0 & 0 & 0 \end{pmatrix} \end{array}, \qquad (3)$$

respectively. In the Fourier-transformed representation, we have the tight binding Hamiltonian

$$H_t = \sum_{\vec{k},\sigma,\gamma\gamma'} \varepsilon_{\vec{k}\gamma\gamma'} c^{\dagger}_{\vec{k}\sigma\gamma} c_{\vec{k}\sigma\gamma'}, \qquad (4)$$

with

$$\varepsilon_{\vec{k}} = 2t\begin{bmatrix} 0 & \cos(k_1+k_2) & \cos k_2 \\ \cos(k_1+k_2) & 0 & \cos k_1 \\ \cos k_2 & \cos k_1 & 0 \end{bmatrix}, \qquad (5)$$

where k_1 and k_2 are the component of the wave vector $\vec{k}$ of the triangular lattice spanned by $\vec{a}_1$ and $\vec{a}_2$, respectively. The indices $\gamma, \gamma(= xy, yz, zx)$ and $\sigma(= \uparrow, \downarrow)$ denote the t_{2g} orbitals and electron spin, respectively, and $c^{\dagger}_{\vec{k}\sigma\gamma}$ ($c_{\vec{k}\sigma\gamma'}$) is the creation (annihilation) operator of an electron with k, σ and $\gamma(\gamma')$. Eq. (2) shows the well-known dispersion relation of the Kagomé lattice (see Fig. 3). This means that the Kagomé lattice structure stays in hiding in the triangular lattice of cobalt ions.

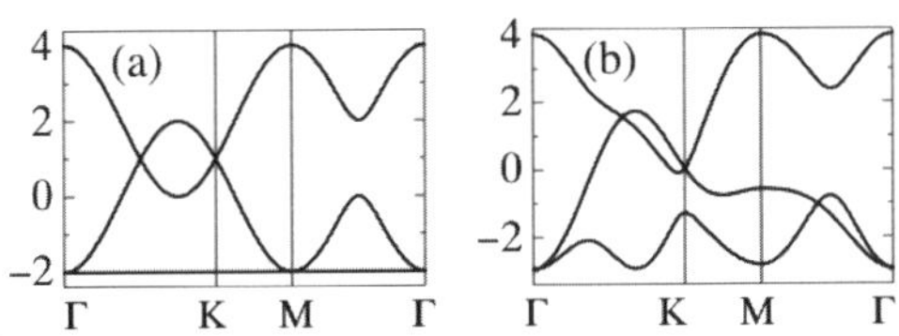

Figure V:1:3. (a) Dispersion relation of Eq. (5) for $t = 1$. (b) Dispersion relation of $\varepsilon_{\vec{k},\gamma,\gamma'}$ for $t = 1$, $t_{dd} = -0.63$, $t_1 = -0.08$ and $t_2/t_1 = -1.5$

Let us discuss how to realize the Kagomé lattice in the motion of an electron given by Eq. (5) An electron in the state $|zx, 1>$ can go to $|yz, 2>$ and $|xy, 3>$. An electron in $|yz, 2>$ can go to $|xy, 12>$ and $|xy,3>$ but cannot go to any orbitals on cobalt 0 due to the symmetry. In this way, an electron starting from $|zx, 1>$ propagates through the t_{2g} orbitals on cobalt ions, 1 ~ 12, and thus the trace of the motion forms a Kagomé lattice (see Fig. 4(a)). The triangle made of the states $|zx, 1>$, $|yz, 2>$ and $|xy, 3>$ is an elementary unit of the Kagomé lattice. Therefore, the energy scheme of the triangle determines that at (0,0) in the k space. The eigenstates of the triangle are

$$(|xy,3>+|yz,2>+|zx,1>)/\sqrt{3} \qquad (6)$$

with the eigenvalue $2t$ and

$$(|xy,3> + e^{\pm i\frac{2\pi}{3}}|yz,2> + e^{\pm i\frac{4\pi}{3}}|zx,1>)/\sqrt{3} \qquad (7)$$

with $-t$. The eigenstates correspond to the a_{1g} and e'_g symmetries, respectively. Note that they are completely different from the states Eqs. (1) and (2). The eigenstates in Eqs. (6) and (7) lie on the top and bottom of the band, respectively. This is a character of the Kagomé lattice structure.

When an electron propagates starting from $|yz, 1>$, the trace can form another Kagomé lattice which is drawn by black triangles in Fig. 4(b). Following the procedure, we obtain four Kagomé lattices as shown in Fig. 4(b). Because the unit cell of the Kagomé lattice is four times as large as that of the triangular one, the four Kagomé lattices complete the Hilbert space.

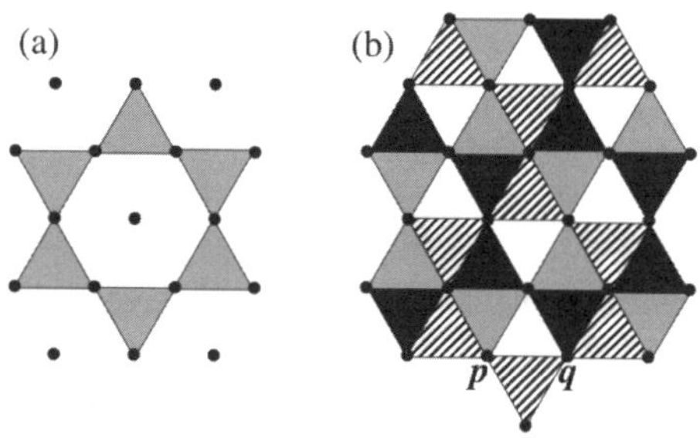

Figure V:1:4. Kagomé lattice in the triangular lattice of cobalt ions. (a) Solid circles indicate the cobalt ions in Fig. 1(b). Gray triangles form a Kagomé lattice which is made by a trace of the travel of an electron starting from $|zx,1>$ (see text). (b) Layout of the four (gray, black, hatched and white) Kagomé lattices

In the CoO_2 layer where the CoO_6 octahedra share the edges as shown in Fig 1, the hopping integral between adjacent 3d orbitals should also be taken into account to analyze the band structure. Between the cobalt ions 1 and 2, the leading term of the hopping matrix element (t_{dd}) occurs between $|xy, 1>$ and $|xy, 2>$. The sign of the hopping matrix element t_{dd} is negative due to the configuration of the orbitals on the hexagonal CoO_2 layer. Although there exist the other hopping matrix elements between the ions 1 and 2, e.g., between $|zx, 1>$ and $|yz, 2>$, their magnitude may be much smaller than the leading term. The hopping between xy orbitals forms a one-dimensional chain along a_1-axis. In the same way, the hopping between yz (zx) orbitals forms another chain along a_2-axis (the direction of $\vec{a}_1 + \vec{a}_2$). Consequently, the hopping matrix in the Fourier-transformed expression is diagonal, i.e., (xy, xy), (yz, yz) and (zx, zx) components are written as $2t_{dd}\cos(k_1)$, $2t_{dd}\cos(k_2)$, $2t_{dd}\cos(k_1+k_2)$, respectively. Note that the hopping matrix does not give the energy-level splitting at (0,0) in the k space.

For more detailed analysis, we introduce the effect of the hopping integral of $2p$ orbitals between neighboring oxygen ions. Let us consider the configuration of $2p$ orbitals on the oxygen ions labeled i ~ vi in Fig. 5 where the relation between xyz and a_1a_2c coordinate systems are the same as that in Fig. 1.

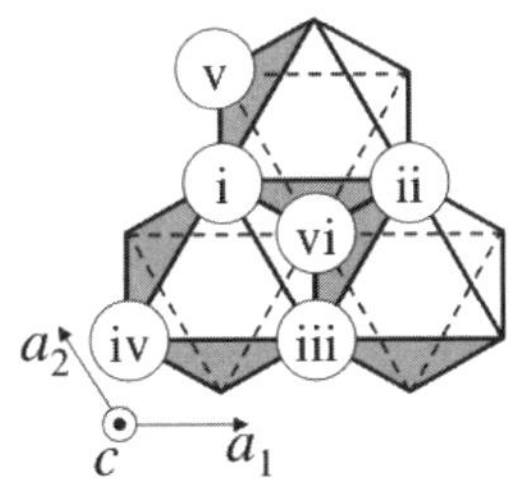

Figure V:1:5. Triangular lattices of oxygen ions in CoO_2 layer. Cobalt ions are not drawn. i ~ iv (v and vi) are on the upper (lower) triangular lattice of oxygen ions

The $2p_z$ orbitals on i, ii and v are on an xy-plane, but those on i and iii are not. Using the table by Slater and Koster [29], we find two kinds of hopping integrals as follows: $t_{pp,1} = V_{pp\pi}$ for the neighboring pair of oxygen ions, (i,ii) and (i,v), and $t_{pp,2} = (1/2)(V_{pp\sigma} + V_{pp\pi})$ for (i,iii), (i,iv) and (i,vi), where $V_{pp\sigma}/V_{pp\pi} = 4$, $t_{pp,2}/t_{pp,1} = -1.5$ and $t_{pp,1} < 0$. Following the procedure, all of the hopping integrals of $2p$ orbitals between neighboring oxygen ions are expressed as $t_{pp,1}$ and $t_{pp,2}$, which lead to the hopping matrix elements, t_1 and t_2, of the $3d$ electron to the second nearest neighbors: $t_n \sim (t_{pd})^2 t_{pp,n}/(\Delta)^2$ where $n = 1$ or 2. As a result, we obtain the hopping matrix $E_{\vec{k},\gamma,\gamma'}$ of the CoO_2 layer:

$$
\begin{aligned}
E_{\vec{k},\gamma,\gamma} = \;& 2(t_{dd}+2t_2)\cos k_a^{\gamma,\gamma} \\
& +2(t_1+2\bar{t}_2)[\cos k_b^{\gamma,\gamma}+\cos(k_a^{\gamma,\gamma}+k_b^{\gamma,\gamma})] \\
& -2t_2[2\cos(2k_a^{\gamma,\gamma}+k_b^{\gamma,\gamma})+\cos(k_a^{\gamma,\gamma}-k_b^{\gamma,\gamma})] \\
E_{\vec{k},\gamma,\gamma'} = \;& 2t\cos k_b^{\gamma,\gamma'}+2t_1\cos 2k_b^{\gamma,\gamma'} \\
& +2t_2[\cos(k_a^{\gamma,\gamma'}+2k_b^{\gamma,\gamma'})+\cos(k_a^{\gamma,\gamma'}-k_b^{\gamma,\gamma'})],
\end{aligned}
$$

where $k_a^{xy,xy}=k_a^{xy,zx}=k_1$, $k_b^{xy,xy}=k_b^{xy,zx}=k_2$, $k_a^{yz,yz}=k_a^{xy,yz}=k_2$, $k_b^{yz,yz}=k_b^{xy,yz}=-(k_1+k_2)$, $k_a^{zx,zx}=k_a^{yz,zx}=-(k_1+k_2)$ and $k_b^{zx,zx}=k_b^{yz,zx}=k_1$, respectively. The dispersion relation captures the band structure calculated by Singh [24]. Although the more parameters may result in the more quantitative agreement with the band structure, it is not the purpose of this paper.

The dispersion relation Fig. 3(b) clearly shows that the upper lying band takes over the nature of the Kagomé lattice structure hidden in the triangular lattice of cobalt ions (see Fig. 4) despite of the presence of t_{dd}, t_1 and t_2. Therefore, it is of crucial importance to study the effect of the Kagomé lattice structure to clarify the electronic state in the CoO_2 layer.

3. Mutual Interactions

Let us discuss the effect of the strong Coulomb interaction in the Kagomé lattice structure shown in Fig 3. The cobalt ions p and q are shared by white and hatched Kagomé lattices. The 3d orbitals $|yz, p>$ and $|zx, q>$ belong to the white Kagomé lattice, whereas $|zx, p>$ and $|yz, q>$ do in the hatched one. The on-site Coulomb interaction brings about the following interaction between p and q when an electron is in the white Kagomé lattice and another is in the hatched one:

$$
\left(\vec{T}_p\cdot\vec{T}_q-\frac{1}{4}\right)\left[J\left(\vec{S}_p\cdot\vec{S}_q+\frac{3}{4}\right)+J'\left(\vec{S}_p\cdot\vec{S}_q-\frac{1}{4}\right)\right],
$$

where $\vec{S}_p$ ($\vec{S}_q$) is the electron spin on p (q). The orbital state on p (q) is described by the pseudo-spin operator $\vec{T}_p(\vec{T}_q)$ with 1/2 in magnitude, i.e., $|yz, p>$ ($|zx, p>$) is the eigenstate of T_p^z with the eigenvalue 1/2 (–1/2) and

$|zx,q\rangle$ ($|yz,q\rangle$) is the eigenstate of T_q^z with the eigenvalue 1/2 (–1/2). The interactions J and J' are expressed as $J = 4t^2/(U' - K)$ and $J' = 4t^2/(U' + K)$ with the Coulomb interaction U' of the inter-orbitals and Hund's rule coupling K. Due to the Hund's rule coupling, $J > J'$, i.e., there exists a ferromagnetic spin coupling with a singlet state of orbitals on the edge shared by the Kagomé lattices.

We propose a fundamental model to study the electronic structure of the CoO_2 layer under the local constraint; $\Sigma_{\sigma\gamma} c_{i\sigma\gamma}^{\dagger} c_{i\sigma\gamma} \geq 5$. The Hamiltonian $H = -H_t + H_J$ with

$$H_J = \sum_{nm} \left(H^{(1)}_{2n\vec{a}_1 + m\vec{a}_2} + H^{(2)}_{m\vec{a}_1 + 2n\vec{a}_2} + H^{(3)}_{2n(\vec{a}_1+\vec{a}_2)+m\vec{a}_2} \right),$$

Table V:1:1. Relation between t_{2g} (xy, yz and zx) orbitals and eigenstates of $T_i^{(I),z}$ with $i = n\vec{a}_1 + m\vec{a}_2$. The letters with (without) the bracket denote the orbitals corresponding to the eigenstates in the case that n + m is odd (even)

eigenvalue	1/2	–1/2
$I = 1$	*yz* (*zx*)	*zx* (*xy*)
$I = 2$	*zx* (*xy*)	*xy* (*zx*)
$I = 3$	*xy* (*yz*)	*yz* (*xy*)

where n and m are integer numbers and

$$H_i^{(I)} = J \sum_{\vec{\delta}(I)} \left(\vec{T}_i^{(I)} \cdot \vec{T}_{i+\vec{\delta}(I)}^{(I)} - \frac{1}{4} n_i^{(I)} n_{i+\vec{\delta}(I)}^{(I)} \right) \times \left(\vec{S}_i \cdot \vec{S}_{i+\vec{\delta}(I)} + \frac{3}{4} n_i^{(I)} n_{i+\vec{\delta}(I)}^{(I)} \right),$$

where I is an index and $\vec{\delta}(1) = \pm\vec{a}_1$, $\vec{\delta}(2) = \pm\vec{a}_2$, and $\vec{\delta}(3) = \pm(\vec{a}_1 + \vec{a}_2)$, in the summation. The orbitals corresponding to the eigenstates of $T_i^{(I),z}$ are summarized in the Ta*f*ble I, and $n_i^{(I)} = n_{i+}^{(I)} + n_{i-}^{(I)}$ where $n_{i\pm}^{(I)}$ is the electron number in the eigenstates of $T_i^{(I),z}$.

This model involves the ingredients for the unique transport and magnetic properties of the CoO_2 layer: A spin-triplet with orbital-singlet is stabilized on a nearest-neighbor cobalt bond. The pairing mechanism acts in a different way from the so-called resonating-valence-bond picture discussed by several authors[18-21], where the key is the singlet state of orbitals. Khaliullin and Maekawa [30] discussed a liquid state of t_{2g} orbitals in a perovskite titanate. In the CoO_2 layer, the resonance and dynamics of the singlet states are developed in the Kagomé lattice but not in the triangular lattice. Note that the orbitals are characterized by four flavors, i.e., the four Kagomé lattices as shown in Fig 4(b), rather than three t_{2g} states. This system has ferromagnetic interaction H_J. Thus, the carrier doping may cause a spin-triplet superconductivity [11-14]. This is based on the Kagomé lattice structure and is different from that on a single band model in a triangular lattice.

The Kagomé lattice involves a triangle as a basic unit. On the triangle, the mechanism by Kumar and Shastry [18, 31] for the anomalous Hall effect will be available and explain the experiments by Wang *et al.*

The triangle gives the a_{1g} state Eq. (6) as the upper lying band in the reciprocal space. In the real space, however, a cobalt ion in the triangle is shared by three Kagomé lattices (see Fig 4), so that the t_{2g} orbitals are identical. The orbital degree of freedom causes the large thermopower [22, 23] at high temperatures.

The Kagomé lattice structure clearly explains the non-symmetric nature of the band structure of the CoO_2 layer. When the effect of the Kagomé lattice becomes dominant, the bottom band, i.e., the flat band as shown in Fig, 3(a) will play a crucial role on the electronic state. Mielke [32] has shown that the flat band with the Coulomb interaction has the ferromagnetic ground state at around half filling. A prospective system for the ferromagnet will be d^1 transition metal oxides, i.e., the layered titanates with iso-structure of the cobalt oxides.

In conclusion, since the cobaltates are in sharp contrast to the high–T_c cuprates in many respects, the study of cobaltates will provide an opportunity to understand the cuprates as well.

ACKNOWLEDGEMENTS

The authors are grateful to K. Tsutsui, T. Tohyama and Y. Ono for useful discussions. This work was supported by Priority-Areas Grants from the Ministry of Education, Science, Culture and Sport of Japan, CREST-JST and NAREGI.

REFERENCES

1. I. Terasaki, Y. Sasago and K. Uchinokura, Phys. Rev. B 56, R12685 (1997).
2. T. Yamamoto, K. Uchinokura, and I. Tsukada, Phys. Rev. B 65, 184434 (2002);T. Yamamoto, I. Tsukada, K. Uchinokura, M. Takagi, T. Tsunobe, M. Ichihara, and K. Kobayashi, Jpn. J. Appl. Phys. 39, L747 (2000); T. Yamamoto, Ph.D. thesis, University of Tokyo, 2001.
3. C. Masset, C. Michel, A. Maignan, M. Hervieu, O. Toulemonde, F. Studer, B. Raveau, and J. Hejtmanek, Phys. Rev. B 62, 166 (2000).
4. Y. Wang, N. S. Rogado, R. J. Cava, and N. P. Ong, Nature 423, 425 (2003).
5. Y. Wang, N. S. Rogado, R. J. Cava, and N. P. Ong, cond-mat/0305455.
6. I. Tsukada, T. Yamamoto, M. Takagi, T. Tsunobe, S. Konno, K. Uchinokura, Phys. Soc. Jpn. 70, 834 (2001).
7. T. Motohashi, R.Ueda, E. Naujalis, T. Tojo, I. Terasaki, T. Atake, M. Karppinen and H. Yamauchi, Phys. Rev. B 67, 64406 (2003).
8. K.Takada, H. Sakurai, E. Takayama-Muromachi, F. Izumi, R.A. Dilanian and T. Sasaki, Nature 422, 53 (2003).
9. R.E. Schaak, T. Klimczuk, M.L. Foo, and R.J. Cava, Nature 424, 527 (2003).
10. Y. Kobayashi, M. Yokoi and M. Sato, J. Phys. Soc. Jpn. 72, 2161 (2003); ibid, 2453 (2003).
11. T. Waki, C. Michioka, M.asaki Kato, K. Yoshimura, K. Takada, H. Sakurai, E. Takayama-Muromachi and T. Sasaki cond-mat/0306036.
12. T. Fujimoto, G. Zheng, Y. Kitaoka, R. L. Meng, J. Cmaidalka, and C.W. Chu, Phys. Rev. Lett. 92, 047004 (2004).
13. K. Ishida, Y. Ihara, Y. Maeno, C. Michioka, M. Kato, K. Yshimura, K. Takada, T. Sasaki, H. Sakurai and E. Takayama-Muromachi, J. Phys. Soc. Jpn. 72, 3041 (2003).
14. W. Higemoto et al., Phys. Rev. B 70,134508 (2004).
15. W. Koshibae and S. Maekawa, Phys. Rev. Lett 91, 257003 (2003).
16. A.Tanaka and X. Hu, Phys. Rev. Lett. 91, 257006 (2003).
17. D. J. Singh, Phys. Rev. B 68, 020503(R) (2003).
18. B. Kumar and B. S. Shastry, Phys. Rev. B 68, 104508 (2003).
19. G.Baskaran, Phys. Rev. Lett. 91, 097003 (2003); cond-mat/0306569.
20. M. Ogata, J. Phys. Soc. Jpn. 72, 1839 (2003).
21. Q.-H. Wang, D.-H. Lee and P. A. Lee, Phys. Rev. B69, 092504 (2004).
22. W. Koshibae, K. Tsutsui and S. Maekawa, Phys. Rev. B 62, 6869 (2000).
23. W. Koshibae and S. Maekawa, Phys. Rev. Lett 87, 236603 (2000).
24. D.J. Singh, Phys. Rev. B 61, 13397 (2000).
25. Y. Miyazaki, M. Onoda, T. Oku, M. Kikuchi, Y. Ishii, Y. Ono, Y. Morii and T. Kajitani, J. Phys. Soc. Jpn. 71, 491 (2002).
26. R. Ishikawa, Y. Ono, Y. Miyazaki and T. Kajitani, Jpn. J. Appl. Phys. 41, L337, (2002).
27. Y. Ono it et al., J. Phys. Soc. Jpn. 70, Suppl. A, 235 (2001); Y. Ono and T. Kajitani, Oxide Thermoelectronics (Research Signpost, Trivandrum, India) 59 (2002).

28. When we adopt a point charge model for the distorted CoO_6 octahedron with the parameters, i.e., the O-Co-O bond angle (98.5°), the ratio of the Co-O bond length and Bohr radius (~4) 10Dq=2.5eV, and the Racah parameter (B = 1065 cm^{-1}) for a cobalt ion (See Y. Tanabe and S. Sugano J. Phys. Soc. Jpn. 9, 766 (1954).), the stabilization energy of the a_{1g} state against the e'_g states is estimated to be ~ 0.025 eV.
29. J. C. Slater and G. F. Koster, Phys. Rev. 94, 1498 (1954).
30. G. Khaliullin and S. Maekawa, Phys. Rev. Lett. 85, 3950 (2000).
31. B. I. Shraiman and R. R. P. Singh, Phys. Rev. Lett. 70, 2004 (1993).
32. A. Mielke, J. Phys. A: Math. Gen. 25, 4335 (1992).

V.2

OXIDE SUPERCONDUCTIVITY

J. D. Dow
Department of Physics, Arizona State University, Tempe, Arizona 85287, U.S.A

Abstract: The trio of ruthenate compounds, doped Sr_2YRuO_6, $GdSr_2Cu_2RuO_8$, and $Gd_{2-z}Ce_zSr_2Cu_2RuO_{10}$ all superconduct in their SrO layers, which is why they have almost the same ≈49 K onset temperatures for superconductivity. The sister compound Ba_2GdRuO_6, whether doped or not, does not superconduct, because the Gd breaks Cooper pairs. These fit in with the superconducting cuprates: the superconducting layers are SrO or BaO, not CuO_2.

Key words: Theories and Models of the Superconducting State; Type II Superconductivity; High-T_c Compounds

1. RUTHENATES AND RUTHENO-CUPRATES

The ruthenates and rutheno-cuprates, including the superconductors (a) Cu-doped Sr_2YRuO_6, (b) $GdSr_2Cu_2RuO_8$, and (c) $Gd_{2-z}Ce_zSr_2Cu_2RuO_{10}$ and the non-superconductor Ba_2GdRuO_6 (whether Cu-doped or not) need to be explained by a successful theory of high-temperature superconductivity. All three of these superconducting compounds begin superconducting within a few degrees of ≈49 K, although full superconductivity, *e.g.*, of $Sr_2YRu_{1-u}Cu_uO_6$, does not become complete until T_c≈23 K, when Ru librations freeze out. Furthermore, whatever we propose for these ruthenates and rutheno-cuprates must also provide a suitable explanation of the cuprates and materials such as $Gd_{2-z}Ce_zCuO_4$ and $YBa_2Cu_3O_7$. The unified explanation which we propose is that the superconductivity is in the SrO or

A. Bianconi (ed.), Symmetry and Heterogeneity in High Temperature Superconductors, 213–216.

the BaO layers (or in the vicinity of interstitial oxygen) of all high-temperature superconductors.

2. DOPED Sr_2YRuO_6 SUPERCONDUCTS

Sr_2YRuO_6 superconducts when it is doped with only 1% Cu on Ru sites [1]. The superconductivity sets in at ≈49 K and becomes rather extensive at ≈30 K (when a spin-glass state appears) but does not become complete until ≈23 K when the Ru librations freeze out.

The interesting thing about the superconductivity is that all the sites are accounted for to better than 1%, and *the material contains no cuprate-planes* (at the >1% level). The material is layered (like most high-temperature superconductors) and so one must decide if the $(SrO)_2$ layers or the doped $YRuO_4$ layers actually carry the supercurrent. Because the Ru has a large magnetic moment, and the SrO has none, we select the SrO layer as the superconducting layer. The data for the material support this choice [2]. Obviously, a cuprate-plane model cannot explain these Cu-doped Sr_2YRuO_6 data, because there are no cuprate-planes in this material. The possibility that the superconductivity does not reside in the cuprate-planes of other high-temperature superconductors, but in the SrO, BaO, or interstitial-oxygen regions, must be considered very seriously.

3. DOPED Ba_2GdRuO_6 DOES NOT SUPERCONDUCT

Neither Cu-doped nor undoped Ba_2GdRuO_6 superconducts, although the material Ba_2GdRuO_6 is virtually the same as Sr_2YRuO_6 (and is in the same class as Sr_2YRuO_6) [3]. The main difference is that Y has L=0 and J=0, while Gd has L=0 and J=7/2. The lack of superconductivity in $Ba_2GdRu_{1-u}Cu_uO_6$ is due to Cooper pair-breaking in the BaO planes by $J \neq 0$ Gd.

4. $GdSr_2Cu2RuO_8$ AND $Gd_{2-z}CezSr_2Cu_2RuO_{10}$

The onset temperatures for superconductivity in the materials $GdSr_2Cu_2RuO_8$ and $Gd_{2-z}Ce_zSr_2Cu_2RuO_{10}$ are also near ≈49 K, and so it appears that doped Sr_2YRuO_6, $GdSr_2Cu_2RuO_8$, and $Gd_{2-z}Ce_zSr_2Cu_2RuO_{10}$ are all members of the same class of superconducting materials, and that they all superconduct in their SrO layers [4].

This viewpoint is bolstered by the facts that both of the ruthenate materials $GdSr_2Cu_2RuO_8$ and $Gd_{2-z}Ce_zSr_2Cu_2RuO_{10}$ show magnetism (associated with their cuprate-planes) at low temperatures, which strongly indicates that their superconductivity is not in their cuprate-planes, but in their SrO layers [5].

5. $Gd_{2-z}Ce_zCuO_4$ DOES NOT SUPERCONDUCT

$Gd_{2-z}Ce_zCuO_4$ does not superconduct, either because the size of Gd is too small or because Gd has $J \neq 0$. The possibility that the problem with the superconductivity is due to Gd being too small can be eliminated by considering $(Gd_{1-u}La_u)_{2-z}Ce_zCuO_4$ which has a bond length that is long enough to superconduct, but does not superconduct. Therefore $Gd_{2-z}Ce_zCuO_4$ does not superconduct because Gd has $J \neq 0$ and is a pair-breaker [6,7] --- much like in Cu-doped Ba_2GdRuO_6.

6. $GdBa_2Cu_3O_7$ DOES SUPERCONDUCT

$GdBa_2Cu_3O_7$ superconducts at around 90 K [8,9]. Since its CuO_2 planes are adjacent to its pair-breaking Gd ions, we can conclude that the superconductivity is not in its CuO_2 planes, but in other layers: the BaO layers are the only such layers that are generally available in other high-temperature superconducting compounds [10].

7. SUMMARY

The ruthenates and rutheno-cuprates are very similar to the cuprates: in both sets of materials the Gd compound does not superconduct, either in $Ba_2GdRu_{1-u}Cu_uO_6$ or in $Gd_{2-z}Ce_zCuO_4$. Also, once the superconductivity is located in the BaO or SrO planes (and not in the cuprate-planes) the superconductivity of $GdSr_2Cu_2RuO_8$ and $GdBa_2Cu_3O_7$ become easy to understand and consistent with the non-superconductivity of $Ba_2GdRu_{1-u}Cu_uO_6$ and $Gd_{2-z}Ce_zCuO_4$.

By placing the superconducting condensate in the SrO or BaO layers, and not in the cuprate-planes, we explain (i) why Ba_2GdRuO_6 does not superconduct, (ii) why $Gd_{2-z}Ce_zCuO_4$ does not superconduct, and (iii) why Sr_2YRuO_6 doped with Cu does superconduct - three facts that have been unexplained by cuprate-plane superconductivity.

REFERENCES

1. S. M. Rao, J. K. Srivastava, H. Y. Tang, D. C. Ling, C. C. Chung, J. L. Yang, S. R. Sheen, and M. K. Wu, J. Crystl. Growth 235, 271 (2002).
2. D. R. Harshman, W. J. Kossler, A. J. Greer, C. E. Stronach, D. R. Noakes, E. Koster, M. K. Wu, F. Z. Chien, H. A. Blackstead, D. B. Pulling, and J. D. Dow, Physica C 364-365, 392 (2001).
3. H. A. Blackstead, J. D. Dow, D. R. Harshman, D. B. Pulling, M. K. Wu, D. Y. Chen, and F. Z. Chien, Solid State Commun. 118, 355 (2001).
4. H. A. Blackstead, J. D. Dow, D. R. Harshman, D. B. Pulling, Z. F. Ren, and D. Z. Wang, Physica C 364-365, 305 (2001).
5. J. D. Dow, H. A. Blackstead, Z. F. Ren, and D. Z. Wang. "Magnetic resonance of Cu and of Gd in insulating $GdSr_2Cu_2NbO_8$ and in superconducting $GdSr_2Cu_2RuO_8$," Submitted.
6. H. A. Blackstead and J. D. Dow, Phys. Rev. B 55, 6605 (1997).
7. H. A. Blackstead and J. D. Dow, Phys. Lett. A 226, 97 (1997).
8. M. K. Wu, J. R. Ashburn, C. J. Torng, P. H. Hor, R. L. Meng, L. Gao, Z. J. Huang, Y. Q. Wang, and C. W. Chu, Phys. Rev. Lett. 58, 908 (1987).
9. Z. Fisk, J. D. Thompson, E. Zirngiebl, J. L. Smith, S.-W. Cheong, Solid State Commun. 62, 743 (1987).
10. D. R. Harshman, W. J. Kossler, X. Wan, A. T. Fiory, A. J. Greer, D. R. Noakes, C. E. Stronach, E. Koster, and J. D. Dow, "Nodeless pairing state in single-crystal $YBa_2Cu_3O_7$", to be published.

V.3

SUPERCONDUCTIVITY VERSUS ANTIFERROMAGNETIC SDW ORDER IN THE CUPRATES AND RELATED SYSTEMS

Inhomogeneities and Electron Correlation

L. S. Mazov

Institute for Physics of Microstructures RAS, Nizhny Novgorod 603600 Russia

Abstract: It is demonstrated that in the cuprates a dynamical, itinerant antiferromagnetic (AF)SDW state (with SDW/CDW stripe structure and d-wave SDW-gap (pseudogap)) is an additional, underlying order for the s-wave Cooper pairing (due to electron-phonon interaction) to appear at higher temperatures. The possible treatment of the nature of AF ordering observed in resistive state in recent neutron scattering experiments on the cuprates is presented. The effects in SC/SDW heterostructures with subnanolayers are discussed.

Key words: Stripe Structure, Spin Density Wave, Pseudogap, Electron-phonon Interaction, s-wave Symmetry, Spin Fluctuations

1. ANTIFERROMAGNETIC SDW PHASE TRANSITION BEFORE SUPERCONDUCTIVITY ONE IN THE CUPRATES

The problem of antiferromagnetic (AF) ordering in applied magnetic field observed at low temperatures in the high-T_c cuprates from neutron scattering [1] (and supporting picture from STM measurements [2]) is intensively discussed now. This AF ordering (spin density wave (SDW)) is

A. Bianconi (ed.), Symmetry and Heterogeneity in High Temperature Superconductors, 217–228.

considered as induced by applied magnetic field, and some theories for explanation of such phenomena was recently forwarded (see, e.g. [3]).

In this work it is demonstrated that itinerant AF SDW in the cuprates is not only competing with superconductivity (SC) order but it is an additional, underlying order which stimulates the SC to appear at higher temperature. The magnetic H-T phase diagram for the cuprates is formed : at low temperatures the SC coexists with SDW/CDW state while above $H_{c2}(T)$-boundary (with increasing T at given H [4] or with increasing H at given T [1,2]) the system enters the pure (non-superconducting) SDW/CDW state.

1.1 The phonon and magnetic contributions in electrical resistivity of the cuprates.

In fig.1 there are presented results of detailed analysis of resistive behavior for YBCO single crystal in the model where total resistivity ρ_{tot} is

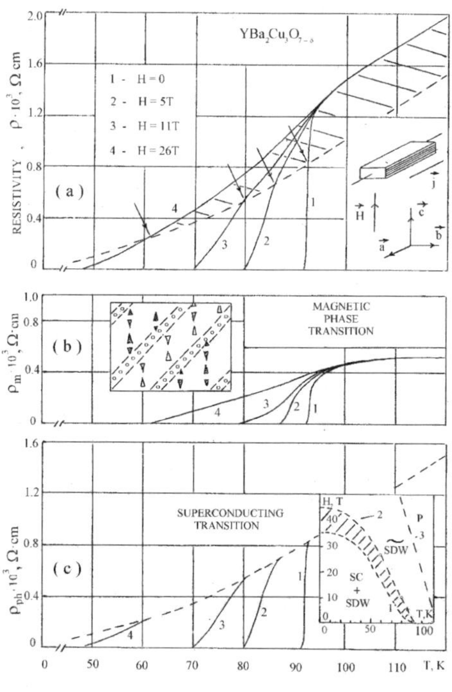

Figure V:3:1. Temperature dependence of the in-plane resistivity for YBCO single crystal [4]

considered as sum of phonon contribution ρ_{ph} and magnetic (due to scattering by AF spin fluctuations) one ρ_m (usual model for magnetic metals) [5]. Phonon contribution $\rho_{ph}(T)$ (dashed curve in fig.1a and 1c) is well fitted by Bloch-Gruneisen curve with Debye temperature $\Theta_D \approx 400K$. The magnetic contribution $\rho_m(T)$ (shaded area in fig.1a) is temperature independent well in the normal state but with decreasing temperature it disappears at the "shoulder" temperature T_k (marked by arrows in fig.1a). Such disappearence is, in fact, evidence for modulated magnetic structure in the system. As it was noted before (see, e.g. [4]), such structure corresponds to SDW/CDW state with stripe structure with alternating spin and charge stripes in CuO_2-plane.

So, from analysis of $\rho_m(T)$- dependence ($\rho_m \approx const$ well in the normal state and $\rho_m = 0$ below shoulder temperature $T_k(H)$) it follows that there is a magnetic (AF SDW) phase transition before SC one in the cuprates (for details, see [4]). With decreasing temperature itinerant SDW/CDW state appears in the HTSC system with onset temperature $T_{SDW} \sim 120-150K$, depending on the family of compounds. At $T = T_{SDW}$ there appear both the stripe structure in CuO_2-plane with alternating spin and charge stripes and SDW-gap at symmetrical parts of the Fermi surface (treated usually as pseudogap (with $d_{x^2-y^2}$ –wave symmetry)). In this temperature region, the spin-fluctuation scattering magnitude is high enough (transverse components in the fluctuation of local spin density, see e.g. [4]). At "shoulder" temperature spin-fluctuation scattering disappears and magnetically ordered state is completed and persists to zero temperature (SC + SDW/CDW state).

It must be noted that the slope of the linear region of the Bloch-Gruneisen curve is here considered as corresponding to the intermediate-temperature linear region ($0.22 < T/\theta_D < 0.43$), which is universal for conventional metals [6]. Note, that the temperature dependence of this linear region of the $\rho_{ph}(T)$ curve for most metals may be described via approximative expression $\rho_{ph}(T)/\rho_1 \approx 0.275(T/\Theta_D) - 0.039$ and it steepness is somewhat higher than that for usually regarded high-temperature region of $\rho_{ph}(T)$ curve with $\rho_{ph}(T)/\rho_1 \approx 0.25(T/\Theta_D)$. (Here ρ_1 is a scaling parameter [4]). It is necessary to emphasize here that the linear dependence of $\rho_{ph}(T)$ for the intermediate-temperature region being extrapolated to $T = 0$ intersects the ρ axis in the point with negative ordinate in contrast to high-temperature one which, in fact, comes through the coordinate origin. It, in fact , denotes that the linear experimental $(\rho(T) \sim T)$ dependence in the normal state, regarded now as characteristic for "good" single crystalline samples of HTSC, is not ruled out the presence of some temperature-independent contribution ρ_m in the total resistivity. It should be underlined that even at $H = 26\ T$ the onset point of SC transition (at $T_k(H)$) also falls in the Bloch-Gruneisen curve (though this curve already deviates from linearity, dashed curve in fig.1c) which fact evidents about applicability of the Bloch-Gruneisen theory to analysis of the HTSC system at $T < \Theta_D/5$, too.

1.2 The magnetic phase H-T diagram for the cuprates.

Such a picture is supported by the magnetic phase *H-T* diagram for HTSC (see, insert in fig. 1c) formed from the data of fig. 1a. The abscisses of "shoulder" points at different values of applied magnetic field H plotted at the (H,T)-plane fall in a straight line (open circles at the lower part of dashed curve 2 in the insert of fig.1c) and considered as corresponding to the onset of the SC transition ($T_c^{onset}(H)$) form, in fact, temperature dependence of the upper critical magnetic field $H_{c2}(T)$. It is essential that there are no problem of the "upward curvature" usually discussed in literature but the dependence of $H_{c2}(T)$ appears to be linear in T near $T_c(0)$ as in the BSC and GL theories. (The "upward curvature" appears in the "$H_{c2}(T)$" dependences obtained, e.g. by "zero-point" methods (curve 1 in the insert of fig.1c). This curve does not correspond to the definition of the term "the upper critical magnetic field" in the type-II superconductors theory and may be considered as boundary between e.g. "vortex lattice" and "vortex liquid" phases). This linearity permits us to estimate the actual value of $H_{c2}(0)$. From the well-known relation: $H_{c2}(T) = H_{c2}(0)(1-(T/T_c)^2)$ it follows that this value is equal to $H_{c2}(0) = 45T$ (see curve 2 (corresponding to this formula) in the inset of fig.1c) what is well coincident with the data of direct measurements in the pulsed magnetic field of $H_{c2}(T \to 0) \approx 42T$ for H along c-axis (see, e.g. [4]). Note here, that from detailed analysis of the data on measurements of the magnetization in the reversible region it follows, in fact the same value for $H_{c2}(0)$ (for details, see [4]). This value of the upper critical magnetic field $H_{c2}(0)$ corresponds to the in-plane coherence length equal to $\xi_{ab} \approx 25\mathring{A}$.

On the other hand, the dependence of $T_k\ (H)$ is, in fact , $T_m^{order}\ (H)$ curve ($\rho_m = 0$), see fig.1b. In this sense, curve 2 and curve 3 (which corresponds to the onset of magnetic phase transition) form the magnetic phase $H - T$ diagram for the given HTCS system. So, below the $T_k\ (H)$ curve the HTSC system appears to be in the coexistence phase: SC and SDW, while above this $H_{c2}\ (T)$ curve the system is in the non-superconducting phase of SDW (modulated magnetic structure which wave number of modulation depends on both H and T (cf. with [4])). Above curve 3 corresponding to the pseudogap (SDW-gap in present treatment) onset temperature $T^*(H)$ (at H > *60 T* at $T \to 0$ (see, e.g. [7]) or at T > *120-150 K* in zero-field case) the system enters the spin-disordered, paramagnetic state.

2. ON THE SYMMETRY OF SC AND SDW/CDW ORDER PARAMETERS IN THE CUPRATES

The above picture is in a good agreement with the theory of itinerant electron systems with interplay between superconductivity and magnetism [8]. In that theory, an itinerant SDW gap may appear at the Fermi surface only before an SC gap, i.e. in the normal state. This SDW gap is highly anisotropic since it is only formed at symmetric parts of the Fermi surface [8] (see, fig.2). Its width Δ_{SDW} being unusually large for an SC gap , well conforms to that for an SDW gap because of inequality $\Delta_{SC} < \Delta_{SDW}$

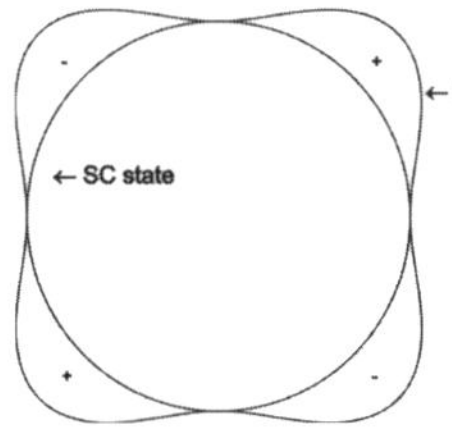

Figure V:3:2. Schematical sketch for coexistence of itinerant SDW and SC states in HTCS

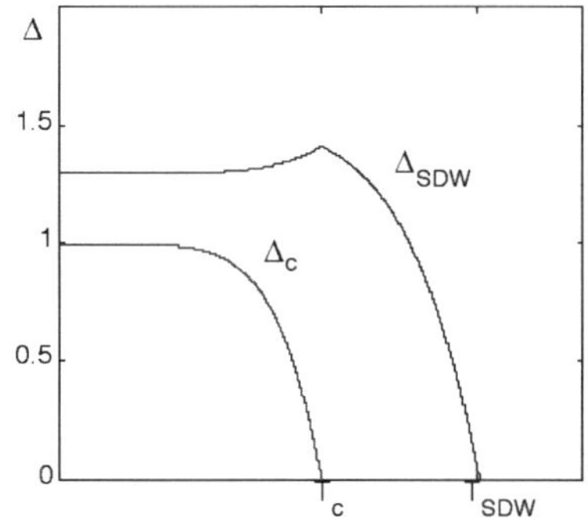

Figure V:3:3. Temperature dependence of SC and SDW order parameters in HTSC cuprates

which is peculiar for the coexistence phase in that model (fig.3). In such a case, the temperature T_c^{onset} (see, fig.1c) may be related to the appearance of the SC gap which begin develop at the Fermi surface only when the transition of the HTSC system to a magnetically- ordered state is over. Then, interrelations : $T_c^{onset} < T_{SDW}^{onset}$ (see, fig.3) and $T_c^{onset} = T_m^{order}$ (see, fig.1a) may be a natural consequence of the equality of magnetic ordering energy ε_m^{order} and the condensation pair one ε_c^{pair} considered as characteristic for such itinerant electron system with an interplay between SC and magnetism. Note here that namely such a picture was recently reported from elastic neutron scattering experiments [9] when in La-based cuprate both the SC and static magnetic order appear at the same temperature.

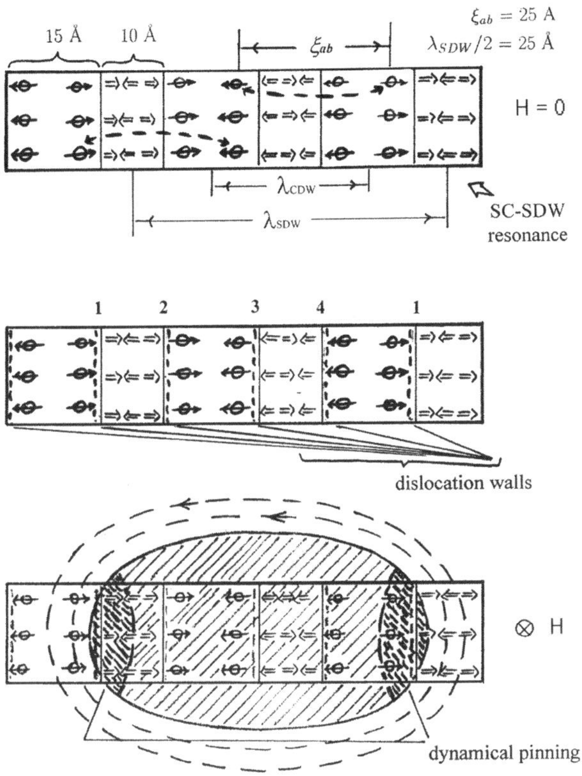

Figure V:3:4. Schematical stripe structure in the CuO_2 -plane

Then, from fig.1c it is seen that SC transition onset points $T_c^{onset}(H)(=T_k(H))$ are at the Bloch-Gruneisen curve (dashed curve). On the other hand, such a picture is characteristic for low temperature superconductors described by s-wave BCS theory. From this analogy it may be concluded that in the cuprates the SC order parameter is of s-wave symmetry, also. Moreover, as seen from magnetic phase *H-T* diagram, the

$H_{c2}(T)$-dependence (formed from SC transition onset points ("shoulder" points)) is linear in T near $T_c(H=0)$, which fact is also characteristic for s-wave BCS and GL theory. Note that such a conclusion about s-wave symmetry of the SC order parameter was supported by a new phase-sensitive test of the order parameter symmetry on the $Bi_2Sr_2CaCuO_{8+\delta}$ bicrystal c-axis twist Josephson junction, the experiment [10] which is considered the strongest one up to date.

The same conclusion follows from analysis of schematical stripe picture in the CuO_2-plane (see fig.4). According to numerical simulation in the Habbard model on a two-dimensional square lattice [11], the arrangement of each spin stripe is antiferromagnetic but two adjacent spin stripes are alternate so that in the CuO_2-plane it is formed SDW with wavelength $\lambda_{SDW} = 50$ Å. The width of stripes corresponds to those measured by Bianconi et al. [12]. This SDW is accompanied by CDW with wavelength equal to one half of that for SDW: $\lambda_{CDW} = \lambda_{SDW}/2 = 25\text{Å}$. Then, as it was obtained from insert in fig.1c, the in-plane coherence length $\xi_{ab} = 25$ Å (see above), so that there is SC-SDW resonance [13].

As it's seen from fig.4 (upper panel), for given stripe structure electrons in charge stripes (arrows with circles) are oriented by spin stripes (double arrows) in such a manner that two electrons at distance equal to coherence length $\xi_{ab} = 25\text{Å}$ can have only opposite directions of spin, which picture is characteristic for the Cooper pairing, i.e. in the s-wave BSC theory. The same result is for other electrons in given charge stripes. This orienting action of spin stripes to stabilize electrons (motion) in charge stripes provide condition for the Cooper pairing at higher temperature. In other words, the SDW ordering is an additional, underlying order for high-T_c superconductivity. In the absence of the SDW order at given temperature thermal fluctuations will destruct the coherent motion of pairing electrons so that pure phonon mechanism (BSC theory) is not able to provide s-wave Cooper pairing at this temperature.

In this concept, the observation of antiferromagnetic (SDW) order in applied magnetic field [1,2] is a direct consequence of restoring (or exposing) of underlying SDW order. In magnetic H-T phase diagram this observation corresponds to motion of (T,H) phase point along H axis (instead of motion along T axis in above resistive measurements (see, fig.1c).

In fig.4 (lower panel) it is schematically presented the structure of the SC vortex in the SDW/CDW + SC state. Since arising of the CDW results in the lattice modulation so that wave of dislocation walls is formed (fig.4 (middle panel)). As known, such dislocation walls are effective centers for pinning of SC vortices. Note that in such a structure every fifth wall is equivalent to first one ($c_{n+4} = c_n$). In this a model a vortex core has AF SDW structure which is also outside a core too. Because of equivalence of c_n and c_{n+4} dislocation walls vortex core becames to be two part in form fluctuating in space (cf. with [14]).

As for the SDW state, then according to the general theory of the SDW (see, e.g. [15]) such modulated state appears as due to the electron-hole pairing in the model of nesting electron and hole Fermi surfaces shifted by vector $\boldsymbol{q} = \boldsymbol{Q}$ in the reciprocal space, i.e. under translation at the nesting vector $\boldsymbol{Q}$ electron and hole surfaces appears to be superimposed. With nesting of the Fermi surfaces the system undergoes the magnetic phase transition in the SDW state with the SDW wave vector $\boldsymbol{Q}$ which appears to be incommensurate with lattice period. If nesting is total (for the whole Fermi surface), then the SDW system appears to be an AFM insulator but when only portion of the surfaces are nested then conductivity of the system is determined by other available for conduction (non-nested) parts of the Fermi surfaces. The condensation of electron-hole pairs with decreasing temperature results in the increase in resistivity but an additional decrease of the scattering rate for noncondensing carriers with decreasing temperature provides a further decrease of the total resistivity. These two opposite processes developing in the SDW system with decreasing temperature may lead to the appearance of a minimum or a shoulder at $\rho(T)$ curve under the SDW phase transition. The description of the SDW system (triplet electron-hole pairing) is in fact similar to that of superconducting system (singlet electron-electron Cooper pairing) in the BCS theory, so that temperature dependence of the SDW and SC gaps are in fact identical (see, fig. 3).

However, the influence of magnetic and nonmagnetic impurities appears to be opposite for the SC and SDW critical temperatures (T_c^{onset} and T_m^{order}, respectively). While influence of nonmagnetic impurities at depression of T_c^{onset} in the SC system is only negligible but in the SDW system normal impurities (say, *Zn*) produce the pair-breaking effect (for electron-hole pairs), the effect well known in excitonic semimetals. Such behavior is in fact analogous with the pair-breaking effect for magnetic impurities in the SC system. And vise versa, it is known that the depression of the ordering temperature in AFM metals with magnetic impurities (say, *Fe*) is only gradual. In the theory of the SDW (see, e.g. [15]) it is supposed that at low substitution level a single *Fe* atom may be coupled to the SDW (cf. with [4]) so that depression of T_c will be not so large. It is interesting to note that namely such (characteristic for the SDW) an effect of magnetic and nonmagnetic impurities is observed in the cuprates (cf. e.g. with [4]).

Then, note also that recently it was introduced a concept of so-called hidden order in the cuprates (see, e.g. [16]). Such an order is attributed to d-density wave (DDW) order. However, in their statement the type of this DDW order is not concrete but it is only considered as competing (not vital) order for SC one, moreover it is considered as corresponding to the superconductivity with $d_{x^2-y^2}$-wave pairing symmetry. As follows from above this concept may be described in terms of the (spin) density wave (S) (DW) with a $d_{x^2-y^2}$-wave symmetry accompanied by (charge) density wave (C) (DW) with $\lambda_{CDW} = \lambda_{SDW}/2$ (see above). The effects of these density waves (DW) are well known and the "hiddenness" of these (DW) in the cuprates may be of natural consequence of the dynamical nature of these (S) and (C) (DW) so that only fast and local probe (including resistivity measurements) permits to detect these (DW) concretely. Moreover, note that the picture of coexistence of triplet electron-hole pairing with SC one and possibility of arising of SC in the spin (DW) state below magnetic (SDW) ordering temperature was discussed before (see, e.g. [4] and fig.1).

It should be noted here that the conclusion about s-wave nature of the SC order parameter is consistent with conclusion about s-wave symmetry of the SC order parameter in the bulk and d-wave symmetry at the surface of the sample of the cuprates [17]. It was noted in [17] that most conclusions about d-wave symmetry was obtained in experiments (e.g. ARPES ones) on the cuprates in which mainly surface phenomena have been used. In this sense, the resistive measurements on the cuprates (see, e.g.[4]) are essentially bulk in the nature. In addition, the electron scattering (in resistivity measurements) is sensitive to the spin disorder in the system (magnetic contribution in the electrical resistivity appears, see Sec.1). Moreover, the electron scattering permits probe not only static magnetic order but dynamical (short-lived) ones because of short characteristic times as compared e.g. with usual neutron scattering.

3. ARTIFICIAL HETEROSTRUCTURES WITH SUBNANOLAYERS TO MODEL THE EFFECTS OF STRIPE STRUCTURE IN THE CUPRATES

Further, if the model is correct then artificial structures with nanolayers of metal and AFM insulator to model high-T_c superconductivity in the CuO_2 layer (fig.4) can be studied. Of course, such a structure is a 3D object

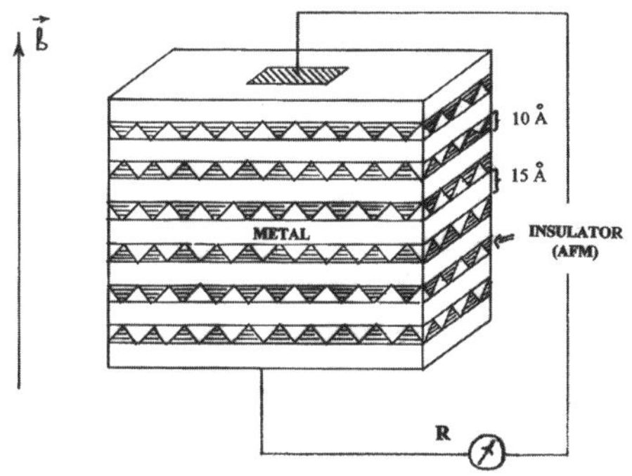

Figure V:3:5. Artificial nanostructure with SDW to model stripe structure in the cuprates

but it has a stripe structure in cross section. The study of temperature dependence of transverse resistivity in a magnetic field can give a contribution to the problem of high-T_c superconductivity. It is essential that such multilayered nanostructures are now available (X –ray mirrors, GMR-structures etc.).

On the other hand, it may be used a secondary effect, namely creation of deformation (distortion) wave in the copper oxide sample, which because of above should lead to the formation of CDW (and hence SDW) state in the volume of a sample with corresponding increase in T_c . It seems likely that namely such methods was used in [18] where thin (15 nm) film of LSCO was grown with block-by-block molecular epitaxy (defect-free growth process) on $SrLaAlO_4$ substrate which lattice period is only somewhat different from that of grown film. Such "incommensurability" results in corresponding deformation (distortion) within thin film and in twofold amplification of T_c ($T_c = 49.1K$) compared with the T_c of bulk compound ($T_c = 25K$). It is essential that in [18] it was noted the change in the $\rho(T)$ dependence in the normal state, characteristic for pseudogap (SDW gap in present treatment) and onset temperature for the pseudogap was noted.

Then, because of dynamical nature of stripe structure in CuO_2 planes there may be used the methods characteristic for solid-state plasma physics, e.g. formation of standing wave along $\vec{b}$ direction with using of, say, spin waves or another ones.

REFERENCES

1. B. Lake et al., Science 291, 1759 (2001) .
2. J.E. Hoffmann et al., Science 297, 1148 (2002).
3. S. Sachdev, Rev. Mod. Phys. 75, 913 (2003).
4. L.S. Mazov, Physics of Met. and Metallogr. S93, 137 (2002); cond-mat/0212128.
5. S.V.Vonsovskii, Magnetism (Nauka, Moscow) 1971.
6. J.M. Ziman, Electrons and Phonons (Oxford Univ. Press, Oxford). 1960.
7. S. Ono et al., Phys.Rev.Lett. 85, 638 (2000).
8. K. Machida, Appl.Phys. A 35, 193, (1984).
9. T. Suzuki, T. Goto, K. Chiba, T. Shinoda, T. Fukase, Phys.Rev. B 57 R3229 (1998).
10. Qiang Li, Y.N.Tsay, M.Suenaga, R.A. Klemm et al. Phys.Rev.Lett. 83, 4160 (1999).
11. M.Kato, K. Machida, H. Nakanishi, M. Fujita, J. Phys. Soc. Japan 59, 1047 (1990).
12. A. Bianconi et al., Phys. Rev. Lett. 76, 3412, (1996).
13. L.S. Mazov, Int. J. Mod. Phys. B14, 3577 (2000).
14. B.W. Hoogenboom, M. Kugler, B. Revaz et al. cond-mat/0002146.
15. V.V. Tugushev, in: Modulated and Localized Structures of the Spin-Density Wave in Itinerant Antiferromagnets in Electronic Phase Transitions (Eds.: W.Hanke and Yu.V.Kopaev, Elsevier, NY/Amsterdam, p.237, 1992).
16. S. Chakravarty, R.B. Laughlin, D.K. Morr, C. Nayak. Phys. Rev. B 63, 094503, (2001).
17. K.-A.Müller, Phil. Mag. Lett. 82, 270 (2002).
18. J.-P. Locquet, J. Perret, J. Fompeirine et al., Nature 394, 453, (1998).

Author Index

Subject Index

Figure Index

Table Index